Psychologists on the March argues that World War II had a profound impact on the modern psychological profession in America. Before the war, psychology was viewed largely as an academic discipline, drawing its ideology and personnel from the laboratory. After the war, it was increasingly seen as a source of theory and practice to deal with mental health issues. With the support of the federal government, the field entered a prolonged period of exponential growth accompanied by major changes in the institutional structure of the field that spread to include the epistemological foundations of psychology.

The book is the first sustained study of this important era in American psychology. Moving back and forth between collective and individual levels of analysis, it provides a narrative that weaves together the internal politics and demography of psychology in relation to the cultural environment. It is based on extensive archival research and includes discussions of the wartime reformation of the American Psychological Association, the role of gender politics, the rise of reflexivity, and the popularization of psychology, among other topics. A series of vignettes focused around the career of Edwin G. Boring, the noted psychologist-historian, provides grounding for the larger story.

James H. Capshew is an associate professor in the Department of History and Philosophy of Science at Indiana University.

Cambridge Studies in the History of Psychology
GENERAL EDITORS: MITCHELL G. ASH AND WILLIAM R. WOODWARD

This series provides a publishing forum for outstanding scholarly work in the history of psychology. The creation of the series reflects a growing concentration in this area on the part of historians and philosophers of science, intellectual and cultural historians, and psychologists interested in historical and theoretical issues.

The series is open both to manuscripts dealing with the history of psychological theory and research and to work focusing on the varied social, cultural, and institutional contexts and impacts of psychology. Writing about psychological thinking and research of any period will be considered. In addition to innovative treatments of traditional topics in the field, the editors particularly welcome work that breaks new ground by offering historical considerations of issues such as the linkages of academic and applied psychology with other fields, for example, psychiatry, anthropology, sociology, and psychoanalysis; international, intercultural, or gender-specific differences in psychological theory and research; or the history of psychological research practices. The series will include both single-authored monographs and occasional coherently defined, rigorously edited essay collections.

Also in the series:

Gestalt Psychology in German Culture, 1890–1967
MITCHELL G. ASH

Constructing the Subject: Historical Origins of Psychological Research
KURT DANZIGER

Changing the Rules: Psychology in the Netherlands, 1900–1985
TRUDY DEHUE

The Professionalization of Psychology in Nazi Germany
ULFRIED GEUTER

Metaphors in the History of Psychology
edited by DAVID E. LEARY

Rebels within the Ranks: Psychologists' Critique of Scientific Authority and Democratic Realities in New Deal America
KATHERINE PANDORA

Inventing Our Selves: Psychology, Power and Personhood
NIKOLAS ROSE

Crowds, Psychology, and Politics, 1871–1899
JAAP VAN GINNEKEN

Measuring Minds: Henry Herbert Goddard and the Origins of American Intelligence Testing
LEILA ZENDERLAND

Psychologists on the March

Subcommittee on Survey and Planning, Emergency Committee in Psychology, National Research Council, 1942. *Top row, left to right:* Richard M. Elliott, Ernest R. Hilgard, Edwin G. Boring. *Bottom row:* Edgar A. Doll, Robert M. Yerkes, Alice I. Bryan, Calvin P. Stone. (Carl R. Rogers is not pictured.) Photograph courtesy of Ernest R. Hilgard.

Psychologists on the March

Science, Practice, and Professional Identity in America, 1929–1969

James H. Capshew

CAMBRIDGE
UNIVERSITY PRESS

PUBLISHED BY THE PRESS SYNDICATE OF THE UNIVERSITY OF CAMBRIDGE
The Pitt Building, Trumpington Street, Cambridge CB2 1RP, United Kingdom

CAMBRIDGE UNIVERSITY PRESS
The Edinburgh Building, Cambridge CB2 2RU, UK   http://www.cup.cam.ac.uk
40 West 20th Street, New York, NY 10011-4211, USA   http://www.cup.org
10 Stamford Road, Oakleigh, Melbourne 3166, Australia

© James H. Capshew 1999

First published 1999

Printed in the United States of America

Typeset in Times Roman 10/12 pt., in QuarkXPress™ [AG]

*A catalog record for this book is available from
the British Library.*

Library of Congress Cataloging-in-Publication Data
Capshew, James H.
Psychologists on the march : science, practice, and professional
identity in America, 1929–1969 / James H. Capshew.
p.   cm. – (Cambridge studies in the history of psychology)
Includes bibliographical references and index.
ISBN 0-521-56267-8 (hardcover). – ISBN 0-521-56585-5 (pbk.)
1. Psychology – United States – History – 20th Century.   I. Title.
II. Series.
BF108.U5C36   1999
150'.973'0904 – dc21   97-47526
CIP

ISBN 0 521 56267 8 hardback
ISBN 0 521 56585 5 paperback

Dedicated to those who showed me
the scholar's way,
at Indiana University,

Thomas F. Gieryn
Eliot Hearst
Herman B Wells

and, at the University of Pennsylvania,

Mark B. Adams
Thomas P. Hughes
Robert E. Kohler
Henrika Kuklick
Russell C. Maulitz
Judith A. McGaw
Charles Rosenberg
Nathan Sivin
Rosemary Stevens
Arnold Thackray
Alexander Vucinich

Like all good teachers
they taught me more than they could ever know.

Stand therefore,
having your loins girt about with truth,
and having on the breastplate of righteousness. . . .
Above all, taking the shield of faith.

Ephesians 6:14, 16

Contents

Acknowledgments

This book got its start as a seminar paper, was expanded into a doctoral dissertation, then underwent extensive revisions, and finally achieved its present form. Along the way I received much help.

Archivists and librarians at the American Philosophical Society, Carnegie-Mellon University, Cornell University, Harvard University, Indiana University, the Library of Congress, the National Academy of Sciences, the National Archives, the Rockefeller Foundation Archives, Stanford University, and Yale University provided access to the documentary sources of this study. A number of psychologists shared their experiences with me, including Neil Bartlett, Alice Bryan, Dorwin Cartwright, Jane Hildreth, Ernest Hilgard, Otto Klineberg, James Grier Miller, Conrad Mueller, Robert Sears, Gertrude Schmeidler, and Dael Wolfle, which added valuable context. Fellow historians, including Mitchell Ash, Philip Bantin, John Burnham, Deborah Coon, Lawrence J. Friedman, Edward Grant, Benjamin Harris, Ellen Herman, Elizabeth Knoll, Henrika Kuklick, Frederic Lieber, Paul Lucas, Jack Pressman, Karen Rader, James Reed, Franz Samelson, Michael Sokal, Richard Westfall, and William Woodward, gave me useful advice.

At a crucial stage in writing I benefited from the editorial suggestions of Janet Donley. My editor at Cambridge University Press, Alex Holzman, has been the model of patience; Helen Wheeler and Barbara Folsom processed the manuscript efficiently.

My students at Indiana University have been a constant source of inspiration, especially Juris Vinters. No less important is the moral support I have received from colleagues in Bloomington and elsewhere. I am grateful for the friendship of Deborah Blankenship, David Carrico, Emily Duncan, Anne Kibbler, Janet Meek, Edith Sarra, Lois Silverman, and a host of others.

My gratitude extends to my partner, Alejandra Laszlo, and to our children, Samantha, Bryna, and Andrew, who have endeavored to teach me the lasting meaning of living in the present.

Candlemas 1998

Abbreviations

ABCP	Advisory Board on Clinical Psychology
ACMH	Army Center of Military History, Washington, D.C.
AGO	Adjutant General's Office Records, National Archives
AIB	Alice I. Bryan Papers, New York City
APA	American Psychological Association Papers, Library of Congress
CML	Chauncey M. Louttit Papers, Yale University Archives
EGB	Edwin G. Boring Papers, Harvard University Archives
GWA	Gordon W. Allport Papers, Harvard University Archives
JGM	James Grier Miller Papers, La Jolla, California
NCWP	National Council of Women Psychologists
PAL	Psycho-Acoustic Laboratory Records, Harvard University Archives
RMY	Robert M. Yerkes Papers, Yale University Archives
RRS	Robert R. Sears Papers, Stanford University Archives
WVB	Walter V. Bingham Papers, Carnegie-Mellon University Archives

Introduction: The Psychologists' War

At the turn of the twentieth century America's foremost psychologist held out the hope that perhaps the "moral equivalent of war" could be found to redirect human energy away from destructive purposes into more positive channels. William James died in 1910 and thus was spared the experiencing of two world wars that gave the lie to the dream he so eloquently expressed. One can only wonder what his reaction would have been to the fact that those global conflicts were not only fought with the aid of tools provided by his fellow psychologists but that they stimulated and shaped the discipline he did so much to establish.[1]

After modest but steady growth from the late nineteenth century, when it was first introduced into American colleges and universities, the field of psychology boomed after the First World War. Between 1919 and 1939 the number of psychologists grew tenfold, from approximately three hundred to three thousand professionals. The Second World War had an even more dramatic and enduring impact. The psychology community had expanded by another order of magnitude by 1970, with more than thirty thousand professionals registered as members of the American Psychological Association. In 1995, fifty years after the end of World War II, the number of psychologists in the United States was approaching a quarter of a million.[2]

This growth transformed psychology from an emerging academic specialty into a mammoth technoscientific profession in less than a century. It increased the proportion of psychologists in the U.S. population from a mere trace to about one to every thousand inhabitants of the country. In today's world it is difficult to find a person whose life has not been touched, directly or indirectly, by a psychologist. The same could be said for other science-based professions, such as chemistry or

1 William James, *The Varieties of Religious Experience* (New York: Penguin, 1985; orig. pub. 1902), p. 367.
2 U.S. Census, 1990.

1

medicine, but psychology has been unique in the pace and extent of its growth since the middle of the century. In the wake of World War II, American society witnessed the creation of a virtually new mental health profession – clinical psychology – that owed its remarkable expansion to its close identification with the aims and practices of scientific research in psychology. Hence the scientific and professional dimensions of psychology developed in tandem in the United States.

It is perhaps not surprising that the world wars strongly affected psychology in America. After all, they were events that reshaped modern society around the globe, with far-reaching cultural as well as political effects.[3] Their impact on science was profound. The First World War has been called "the chemists' war" because of the introduction of poison gas as a potent new weapon of mass destruction. Deployed by both sides in the conflict, it raised important ethical issues about the use of scientific knowledge in modern warfare.[4] It also contributed, especially in the United States, to increased federal support of research and development and to new organizations that bound the scientific community in service to the state. The mobilization of American scientists for World War II was even more fateful. The development of the atomic bomb meant that this conflict would go down in history as "the physicists' war."[5] But the bomb had significant psychological fallout as well. Postwar politics ensured the continued reliance on scientific research and technological innovation as the basis for national security. For nearly a quarter of a century the U.S. scientific community prospered as never before, as its fortunes became inextricably tied to the Cold War.[6]

It was not until the war in Vietnam escalated during the late 1960s that the intimate relationship between science and the federal government was called into question by scientists and others. In the ensuing years, federal research and development budgets stopped growing as rapidly as they had been, and in some cases

3 "The Second World War is the largest single event in human history, fought across six of the world's seven continents and all its oceans. It killed fifty million human beings, left hundreds of millions of others wounded in mind or body and materially devastated much of the heartland of civilization." John Keegan, *The Second World War* (Toronto: Key Porter Books, 1989), p. 5.

4 See Hugh R. Slotten, "Humane Chemistry or Scientific Barbarism? American Responses to World War I Poison Gas, 1915–1930," *Journal of American History,* 1990, *77:*476–498.

5 See Daniel J. Kevles, *The Physicists: The History of a Scientific Community in Modern America* (New York: Knopf, 1978).

6 See Paul Forman, "Behind Quantum Electronics: National Security as Basis for Physical Research in the United States, 1940–1960," *Historical Studies in the Physical Sciences,* 1987, *18:*149–229; Stuart W. Leslie, *The Cold War and American Science: The Military-Industrial-Academic Complex at MIT and Stanford* (New York: Columbia University Press, 1993).

declined, whereas the number of physical scientists continued to increase, thanks to a productive Ph.D. machine. The biomedical sciences were an exception, however, because of their connections to an expanding health care industry. The social and behavioral sciences, in further contrast, remained largely tied to the fortunes of the academic market.

Psychology continued to burgeon throughout the postwar period, as it proved itself to be enormously versatile and adaptive. As a highly academic subject, it benefited from the rise of the American research university and the expansion of higher education more generally.[7] It had tremendous popular appeal as well, and there was a seemingly insatiable public appetite for psychological facts and techniques. The use of mental tests during the First World War demonstrated how a science-based technology could serve the interests of the government while simultaneously being turned toward professional self-advancement. The Second World War further reinforced that lesson, as psychologists touted themselves as indispensable experts on the "human factor" and focused their technoscientific tools on a vast array of wartime problems.

After that war, psychology capitalized on its manifold identity as a natural science, a social science, and a mental health profession and took advantage of multiple sources of support. Although retaining a strong academic foundation, it was able to develop new occupational niches outside the university in the delivery of mental health services. Scientific research and professional practice flourished together in a symbiotic relationship, as traditional distinctions between pure and applied psychology broke down in the face of demands for social utility.

William James, as a scientist and a scholar, was deeply committed to humanistic ideals. In his efforts to construct a naturalistic science of human nature, he looked first and foremost at evidence derived from personal experience, his own as well as that of others. He also viewed the psychological laboratory as a useful source of information and was among the first Americans to appreciate the scientific revolution that the experimental approach to the study of the mind had wrought in Europe in the late nineteenth century. In his role as a Harvard professor, he made sure that the university was well equipped for experimental research, although he himself lacked the patience and dexterity for such work.

After James published *The Principles of Psychology* in 1890, which established his reputation as America's preeminent psychologist, he became increasingly dissatisfied with the narrow, technoscientific turn the "new psychology" was taking in the United States. In their rush to prove that their work was scientific, he thought

7 See Richard M. Freeland, *Academia's Golden Age: Universities in Massachusetts, 1945–1970* (New York: Oxford University Press, 1992); Roger L. Geiger, *Research and Relevant Knowledge: American Research Universities since World War II* (New York: Oxford University Press, 1993).

that psychologists had become preoccupied with methodological rigor and were losing sight of the larger goals of knowledge. James was more interested in describing the adaptive functions of consciousness than in delineating the structure and contents of the mind and found that philosophy provided a more congenial and supportive environment for the development of his ideas about pragmatism.[8]

James's doubts notwithstanding, laboratory work in psychology burgeoned. By 1900 more than forty academic laboratories had been started in America, and the new psychology was being rapidly professionalized. The laboratory assumed an almost religious significance in this period as psychologists fervently asserted the claims of science in the study of human nature. In the battle for truth, superstition and ignorance were the enemies, to be vanquished by the sharp sword of experiment. Whatever their differences, psychologists shared a fundamental faith in the power of science not only as a means to acquire valid knowledge but as a way of organizing the world.[9]

When James died in 1910 there were only a few hundred individuals professing psychology in the entire United States. But their technoscientific crusade had already begun and was gathering momentum. Psychology, as well as science more generally, received an enormous boost from the global conflict that erupted into the First World War, which demonstrated the continuing failure of Western civilization to discover a moral equivalent to war. Not only did the war generate a myriad of human problems amenable to professional intervention, it provided a rationale for solving them. As citizens and as scientists, psychologists could do no less than contribute their expertise toward the national defense.

But the merging of public duty and professional self-interest had some significant consequences when many psychologists left their laboratories behind and made their work practical. They conducted mental tests on an unprecedented scale, measuring the intelligence, aptitudes, and skills of huge numbers of armed forces personnel. The success of psychology in the war effort thus rested mainly on the correlational approach of the mental testers, not on the experimental protocols of the laboratory workers. Within the profession, the tension between the respective pursuits of knowledge for its own sake and for the sake of utility was often couched in terms of "pure" versus "applied" psychology, with the experimentalists claiming the high ground of science and disparaging work motivated by less lofty goals, despite the fact that the practical work of mental testing had "put psychology on the map" in the first place.

8 See Daniel W. Bjork, *William James: The Center of His Vision* (New York: Columbia University Press, 1988).
9 Ernest R. Hilgard, *Psychology in America: A Historical Survey* (San Diego, Calif.: Harcourt Brace Jovanovich, 1987), p. 31. See also John M. O'Donnell, *The Origins of Behaviorism: American Psychology, 1870–1920* (New York: New York University Press, 1985); James H. Capshew, "Psychologists on Site: A Reconnaissance of the Historiography of the Laboratory," *American Psychologist,* 1992, *47:* 132–142.

Following the war the American psychology community grew rapidly, numbering perhaps a thousand professionals by 1929. As the academic labor market became saturated and the Great Depression limited other employment opportunities, the psychology community became increasingly fragmented by the continued influx of new advanced-degree holders. Although behaviorism provided some degree of ideological unity, the discipline was riven with debates over the theory, methods, and goals of psychology. The role of the psychologist was also being redefined, as practitioners sought to enlarge their sphere of influence in a highly academic profession.

What one observer called "the impending dismemberment of psychology" along ideological, organizational, and generational lines was averted by the Second World War.[10] Patriotic duty again became synonymous with professional service, as psychologists suspended their internecine warfare for the duration and once more marched in unison for the common defense of their country. Indeed, the experience of World War II transformed America's psychologists individually and collectively. The profession mobilized itself rapidly and thoroughly, finding places for psychology in a vast assortment of wartime agencies and projects, both military and civilian. Personnel administration, morale, propaganda, man–machine engineering, and mental health were among the areas in which psychologists came to be deployed. Widespread mobilization caused a radical short-term occupational shift that had long-term consequences. Academic psychologists who had spent their time in the laboratory and the classroom with rats and college students suddenly found themselves dealing with real-world problems involving adult human subjects. Serving as technoscientific experts only reinforced psychologists' already strong sense of social purpose. And the dropping of the atomic bomb confirmed their conviction that they had an essential contribution to make toward viable human relations.

This study explores the complex relations that existed among professionalism, science, and ideology in American psychology during the middle of the twentieth century. World War II stands as a watershed in the development of psychological science, when social forces and intellectual trends engendered a major shift in the role of the psychologist. As the professional self-image of the psychologist was transformed, public attitudes and expectations about psychology also changed.

Like all professionals, psychologists sought to define their occupational identity, to control the conditions of their employment, to exercise a monopoly over

10 Heinrich Klüver, "Psychology at the Beginning of World War II: Meditations on the Impending Dismemberment of Psychology Written in 1942," *Journal of Psychology,* 1949, *28*:383–410. See also Franz Samelson, "The 'Impending Dismemberment of Psychology' and Its Miraculous Rescue, 1930–45," unpublished paper delivered at annual meeting of the Cheiron Society, 1988.

their special expertise, and to regulate their ethical standards. As scientists, they wanted to achieve the epistemological goals of certified knowledge and attain the status accorded to members of more established disciplines. What brought these two realms of aspiration together was an ideology founded on the prospects for scientific understanding of human nature and conduct, and harnessed to the project of cultivating a productive symbiosis between the pursuit of psychological knowledge and its applications.

Both world wars were instrumental in the development of psychology as a technoscientific profession. They did more than "prove" that psychology was of practical use; they brought new opportunities and increased resources, moral as well as financial, to the community. They also helped psychologists to think of themselves not only as professional scientists but also as scientific professionals. In many respects the First World War was a dress rehearsal for the Second, as a close relationship with the federal government was reestablished and several key leaders in psychology reprised their roles. But the growth of the psychology community between the wars, along with the greater scale and intensity of World War II, led to correspondingly larger changes in the shape and direction of the discipline.

Psychologists exploited new opportunities to expand their professional roles in the wake of World War II and the field grew at an exponential rate. With secure institutional support and ample resources, psychology exploded intellectually. The introduction of new concepts, techniques, and instruments, some indigenous to psychology and some borrowed from other disciplines, created a dizzying array of research specialties ranging from the biological bases of behavior to the social psychology of small groups. It seemed as if wherever human nature was involved, psychologists were there to analyze and prescribe.

Psychology achieved its identity as a coherent field largely through organizational means, such as standardized college curricula and certified graduate training programs. On an intellectual level, psychology remained heterogeneous and multiparadigmatic. But a deep change was occurring in the epistemological foundations of the discipline as psychologists embraced the reflexive implications of their work. This allowed for the incorporation of personal experience into the realm of science and provided the key to the productive tension between scientific research and professional practice that characterized postwar psychology in America. The result was the creation of a technoscientific system that churned out new knowledge and new psychologists at a prodigious rate and provided the basis for the proliferation of psychological ideas and techniques in American society.

The war launched American psychologists on a highly visible trajectory as cultural authorities on the human psyche. In the postwar decades, scientific psychology flourished as it spawned a huge "helping profession" dedicated to solving all manner of psychological problems. The work of psychologists as diverse as B. F. Skinner, Joyce Brothers, Carl Rogers, and Timothy Leary became well known as they took their technoscientific gospel directly to the public. After the war, psychology was among the fastest-growing fields in America, as science and engineering enjoyed unprecedented support that fueled spectacular growth for nearly a quarter-century.

Just as protecting national security provided the rationale and context for research in physics in postwar America, so the achievement of human potential served as the basis for the production of scientific knowledge in psychology. For psychologists as well as physicists, World War II provided the opportunity and the motivation for a new relationship between their work and American society. By making new ideological and material resources available, the war enabled such technoscientific professions to expand their spheres of influence and shape postwar life in unprecedented ways.

This book is organized as an examination of the causes and consequences of growth in psychology. Fundamentally, it seeks answers to why the field expanded so quickly in the middle years of the twentieth century. It takes as its starting point the history and culture of the psychology profession and tries to ground the analysis in the activities of flesh-and-blood psychologists. I have sought to interpret psychologists' individual or collective behavior as emblematic, providing clues to the larger context of American culture.

The book is divided into several sections. The main narrative begins with three chapters on the legacy of the First World War and its influence as psychologists mobilized for World War II. Chapters 4–7 deal with the activities of psychologists as they sought to make their expertise available during the war itself. The postwar environment for psychology is treated in Chapters 8 and 9. The final two chapters explore the epistemological and cultural consequences of psychologists' wartime experiences. Each section of the narrative is set off by a short interlude that focuses attention on the career of Edwin Boring. These vignettes are designed to show how the collective transformation of the field was reflected in the life of a single individual. In attempting to relate the microcosm of the profession to the macrocosm of society, the book borrows from the traditions of disciplinary history and cultural studies.[11] What follows is an exploration of "the psychologists' war" and its enduring disciplinary, professional, and cultural impact.

11 See, for example: Mitchell G. Ash, *Gestalt Psychology in German Culture, 1890–1967: Holism and the Quest for Objectivity* (New York: Cambridge University Press, 1995); John C. Burnham, *Paths into American Culture: Psychology, Medicine, and Morals* (Philadelphia: Temple University Press, 1988); Deborah J. Coon, "Standardizing the Subject: Experimental Psychologists, Introspection, and the Quest for a Technoscientific Ideal," *Technology and Culture,* 1993, *34:*757–783; Kurt Danziger, *Constructing the Subject: Historical Origins of Psychological Research* (Cambridge: Cambridge University Press, 1990); Gerald N. Grob, *From Asylum to Community: Mental Health Policy in Modern America* (Princeton, N.J.: Princeton University Press, 1991); Ulfried

Geuter, *The Professionalization of Psychology in Nazi Germany* (New York: Cambridge University Press, 1992); Ellen Herman, *The Romance of American Psychology: Political Culture in the Age of Experts* (Berkeley/Los Angeles: University of California Press, 1995); Bruno Latour, *Science in Action* (Cambridge, Mass.: Harvard University Press, 1987); John M. O'Donnell, *The Origins of Behaviorism: American Psychology, 1870–1920* (New York: New York University Press, 1985).

Interlude I

Whether by choice or by chance, psychologists at Harvard University have been making headlines for more than a century. Beginning with William James, author of *The Principles of Psychology* (1890), America's oldest academic institution has harbored a succession of prominent and controversial figures. James, credited with starting the first psychological laboratory in the country, lost faith in the emerging discipline as it became preoccupied with the experimental method in the production of new knowledge. Before abandoning psychology for philosophy, however, he arranged for a major expansion of the Harvard laboratory, including the importation of experimental Hugo Münsterberg from Germany to direct it. Münsterberg, who was hired in 1892 as a representative of the pure research ideal, proved to have equally strong convictions about the importance of applied psychology and became a highly visible proponent of the social utility of psychological knowledge. During the First World War, Münsterberg's outspoken views on the superiority of German culture made him a lightning rod for criticism, and his loyalty was called into question. In 1916, a few months before the United States entered the war, a stroke killed him while he was delivering a public lecture.[1]

Münsterberg's eventual successor was Edwin G. Boring, who became director of the laboratory in 1924 and set out to restore the primacy of experimental work in psychology, at Harvard and elsewhere. Like his predecessors, he became one of America's best-known psychologists. Teacher, historian, and editor, Boring used his Harvard professorship as a bully pulpit to educate one and all about the virtues and vices of his chosen discipline. Nothing if not passionate about establishing psychology on a secure scientific basis, Boring was equally concerned about the professional identity of the field. Over the course of his career he witnessed the spectacular growth of the American psychology community from a small band of

---

1 Matthew Hale, Jr., *Human Science and Social Order: Hugo Münsterberg and the Origins of Applied Psychology* (Philadelphia: Temple University Press, 1982).

9

academic specialists into a virtual army of scientific professionals and allied technical workers. He also lived through several paradigm shifts, as psychologists abandoned the introspective study of the mind in favor of behaviorism and then, later on, joined the "cognitive revolution" that transformed the behavioral sciences.

Boring's career spanned nearly six decades, including twenty-five years as director of the Harvard psychological laboratory. He had begun as a prolific and versatile experimenter and then turned to historical scholarship. His reputation rested on his massive and erudite book, *A History of Experimental Psychology,* published in 1929. Deeply involved in professional affairs, Boring served as a leader in the country's main professional organization, the American Psychological Association, and was an active participant in the elite Society of Experimental Psychologists. In all of his myriad efforts on behalf of psychology, Boring aspired to be what he called a "commanding servant" – one who could exercise benevolent authority because of a total identification with the interests of the whole group.[2]

By the end of his life Boring had become one of the "great men" of contemporary American psychology. He was known throughout the professional community and by the public at large for his pithy pronouncements on all matters psychological. Having erased the boundary between his private life and his public role, Boring virtually embodied the ongoing search for professional identity that had engaged psychologists, both individually and collectively, during the twentieth century. In the vignettes that follow, Boring's life and work will be used to reveal in microcosm the forces and events that shaped American psychology.[3]

In the late spring of 1914, Edwin Garrigues Boring was awarded a doctorate from Cornell University in the young science of psychology. Fewer than 350 students had ever received such a degree in the United States. Twenty-eight years old, the ambitious Quaker-Moravian from Philadelphia had already begun to prove himself as a productive experimenter. He had been trained by Edward Bradford Titchener, the prominent leader of the structuralist school of thought and a leading advocate of the academic ideal of pure research.

Boring first encountered Titchener in 1905 when he was an undergraduate engineering student at Cornell. Remembering the lectures in elementary psychology as "magic," Boring went on to complete his engineering degree in 1908.[4] After a year of work in a steel mill and another as a high school teacher, Boring returned to Cornell intending to pursue a master's degree in physics. But the encouragement

2 Edwin G. Boring, in *A History of Psychology in Autobiography,* vol. 4 (Worcester, Mass.: Clark University Press, 1952), pp. 27–52, quote on p. 51.

3 As one of his students noted, "Boring's identity was his bedevilment." Saul Rosenzweig, "E. G. Boring and the *Zeitgeist: Eruditione Gesta Beavit," Journal of Psychology,* 1970, *75:*59–71, quote on p. 69.

4 Edwin G. Boring, *Psychologist at Large* (New York: Basic Books, 1961), p. 18.

he received from psychology professor Madison Bentley and the lure of Titchener's lectures seduced him into psychology, where he soon found his vocation.

As Titchener's student, Boring was bequeathed a distinguished academic genealogy. Titchener, who was British by birth and an Oxford graduate, had studied under Wilhelm Wundt in Leipzig, Germany, and earned his Ph.D. there in 1892. Wundt was a major figure in the development of *physiologische Psychologie* – the so-called new psychology – that addressed the perennial philosophical problems of mind and consciousness using new scientific tools. Titchener took one aspect of his mentor's complex system, the method of experimental introspection, and made it the linchpin of his own. By virtue of his translations of Wundt's voluminous works and his own system-building efforts, Titchener became identified as Wundt's intellectual heir after moving to Cornell in 1892.

Under Titchener's tutelage Boring obtained a rigorous and thorough education in science. He became well acquainted with the German-language literature of psychology and took to heart Titchener's dictum "that the historical approach to understanding of scientific fact is what differentiates the scholar in science from the mere experimenter."[5] As a novice instructor at Cornell, Boring was responsible for teaching the introductory psychology course that Titchener had developed into a comprehensive survey of the field.

Titchener embodied the stereotype of the imperious German *Professor.* Autocratic and dictatorial, he commanded allegiance from his disciples. Boring, whom Titchener considered his best student, submitted willingly to such discipline, even to the point of returning early from his honeymoon when the master summoned him back to the university. His new bride, Lucy Day Boring, no doubt understood, having earned her own doctorate under Titchener in 1912.[6]

During his graduate school days Boring gained broad experience as a researcher. In addition to his focus on the core areas of psychophysics and physiological psychology, he conducted studies in animal, educational, and abnormal psychology, as well as in nerve physiology. His dissertation on the sensations of the alimentary canal was based on investigations of his own body via a tube inserted into his stomach.[7] Although Boring worked on an eclectic array of topics, the laboratory served as his ideological lodestar, signifying his unwavering commitment to experimentation as the basis for scientific authority in psychology.

After completing his doctorate Boring remained at Cornell as an instructor. Although opportunities for advancement were limited, being part of Titchener's inner circle had its compensations. Happily ensconced in the self-proclaimed center

---

5 Ibid., p. 3.

6 Ibid., p. 25. His wife's obituary (1886–1996) mentions that she "married psychology" while pursuing a career as a housewife and mother.

7 For a list of publications from these researches, see Boring, *Psychologist at Large,* pp. 347–348.

of scientific psychology in America, Boring might have continued along in his lab-centered life had the First World War not intervened. After being passed over by the draft on account of his age and two young children, he volunteered for service and was commissioned as a captain in the army medical corps. He reported for duty in February 1918 to Camp Greenleaf, Georgia, where the army's intelligence testing program was being organized. Although the war took him away from Cornell permanently, Boring was never to escape completely from Titchener's hold over his imagination and the heritage of German *Wissenschaft* that he represented.

The First World War marked the beginning of Boring's lifelong quest for professional identity. Leaving the safe confines of the Cornell laboratory for the first time, he encountered psychologists who were bent on making their science immediately useful by means of mental tests. Titchener had always denigrated applied psychology, arguing that technological applications should be pursued only after a firm foundation of scientific knowledge had been constructed. Boring rationalized his involvement on the grounds of patriotic duty, viewing his wartime work as a temporary diversion. But personal contact with the testers changed his attitude:

> Titchener's in-group at Cornell had appreciated mental testers in much the same way that the Crusaders, gathered around Richard Coeur-de-Lion, appreciated Moslems, but this First World War gave me a respect for the testers. I saw clearly that good, honest, intelligent work in any field merits respect and that testers closely resemble the pure experimentalists in habits of work, in enthusiasm, and in thoroughness.[8]

Enormously consequential for Boring were the relationships he established with two slightly older colleagues, Robert M. Yerkes, who headed the army project, and Lewis M. Terman, its research director. Yerkes was an eclectic and ambitious comparative psychologist, dedicated to making psychology the scientific basis of social control and cultural renewal. Terman, through his work on the Stanford-Binet intelligence test, was beginning to make his mark as a leading mental tester. Both were convinced that psychology should be more than an academic profession and were determined to spread their Progressivist faith in science by making psychology practical. Boring, on the other hand, had adopted Titchener's more detached perspective that psychologists should be primarily concerned with the scientific adequacy of their knowledge before turning their work in technological directions. Despite these fundamental differences in outlook, the three men developed deep and enduring professional friendships.

In 1919, after helping Yerkes complete a massive summary report on the World War I testing program for the National Academy of Sciences, Boring took up a professorship at Clark University. There he inherited the psychological laboratory

8 Ibid., p. 31.

started by G. Stanley Hall, who was a year away from his retirement as president of the university. Hall, who had been a student of William James, founded the American Psychological Association in 1892 and served as its first president. Under his leadership, Clark University had enjoyed an impressive reputation in psychology for more than two decades.

After three years at Clark, Boring received the coveted call from Harvard University. Convinced by Titchener that psychology there "needed to be brought clearly into the scientific circle and rescued from the philosophers who still dominated it," Boring accepted the challenge and spent the next decade making a place for experimental research at Harvard. To buttress his arguments about the proper contemporary role of psychological science, he produced *A History of Experimental Psychology*, which traced the Germanic legacy of the field passed down from Fechner, Wundt, and Titchener.[9]

In writing about the history of psychology Boring adopted a personalistic approach that emphasized the intellectual accomplishments of a succession of major contributors to the field. His preoccupation with the "great man" theory of history was strongly rooted in his relationship with Titchener, the self-appointed representative of Wundtian psychology in the United States. When Titchener died in 1927, Boring was in the midst of writing his history. Recalling his reaction, he reported, "it was as if the Ten Commandments had suddenly crumbled." The loyal but conflicted disciple immediately wrote a long obituary of his academic father. At the time he thought he was writing simply to honor Titchener, but, looking back, Boring realized that he was also "moved by the suppressions of seventeen years of ambivalence, enthusiasms mixed with frustrations."[10]

Boring's ambivalence about Titchener revolved around philosophical and methodological issues, in particular the role of introspection as a research technique. Programmatically, however, Boring was completely in accord with Titchener's identification of psychology as a "pure" science and his insistence on the pursuit of psychological knowledge for its own sake, unsullied by the demands of the marketplace for practical information or technological products. Both men believed that a credible applied psychology was not possible until a secure scientific foundation had been established.

Boring concluded his *History* with the declaration that "the application of the experimental method to the problem of mind is the great outstanding event in the history of the study of mind." To those who might disagree with his assessment he offered an ad hominem rebuttal: "the person who doubts that the results have justified

9 Ibid., p. 33; Boring, *A History of Experimental Psychology* (New York: Century, 1929).
10 Edwin G. Boring, "Edward Bradford Titchener: 1867–1927," *American Journal of Psychology,* 1927, *38:*489–506; reprinted with an introductory note in Boring, *Psychologist at Large,* pp. 246–265, quote on p. 247.

the importance that has been attached to the invention of experimental psychology must be either ignorant or influenced by a disappointment that the progress of experimental psychology has not aided in the solution of his own particular problems."[11] Ironically, by writing about the ostensible progress of experimental psychology, Boring was able to find at least a partial solution to the problems he encountered in establishing his own professional identity.

11 Boring, *History of Experimental Psychology,* p. 659.

1

Growing Pains

After the Great War

"Enfin! Enfin en Amérique!" exclaimed Edouard Claparède at the opening session of the Ninth International Congress of Psychology in early September 1929. The French psychologist was welcoming more than eight hundred colleagues from around the world who had gathered at Yale University to participate in the first such international meeting ever held in the United States. From the time of the First International Congress of Psychology that took place in 1889 in Paris, proposals had been made to hold a meeting in America, but political and economic circumstances had intervened. Finally, on the fortieth anniversary of the first congress, delegates from twenty-one foreign countries convened in New Haven, Connecticut. Although most hailed from Europe and the Soviet Union, there were also representatives from China, Japan, India, Australia, New Zealand, Egypt, and Brazil. Outnumbering their foreign guests nearly seven to one, American psychologists were proudly demonstrating the growth and prosperity of the discipline in their country.[1]

The Ninth International Congress signaled that American psychology had come of age. Scientific psychology had begun in Europe during the latter part of the nineteenth century and was soon imported to North America, where it flourished with remarkable vigor. According to the 1929 *Psychological Register*, an international directory, there were more professional psychologists in the United States than in

[1] Edouard Claparède, "Esquisse historique des Congrès Internationaux de Psychologie," in Ninth International Congress of Psychology, *Proceedings and Papers* (Princeton, N.J.: Psychological Review Co., 1930), pp. 33–47, quote on p. 33. See also Carl P. Duncan, "A Note on the 1929 International Congress of Psychology," *Journal of the History of the Behavioral Sciences,* 1980, *16:*1–5; Herbert S. Langfeld, "The Ninth International Congress of Psychology," *Science,* 1929, *70:*364–368; Michael M. Sokal, "James McKeen Cattell and American Psychology in the 1920s," in *Explorations in the History of Psychology in the United States,* ed. J. Brozek (Lewisburg, Pa.: Bucknell University Press, 1984), pp. 273–323; Katherine Adams Williams, "Psychology in 1929 at the International Congress," *Psychological Bulletin,* 1930, *27:*658–663.

the rest of the world combined.[2] The United States also led the world in producing knowledge in psychology, outstripping Germany in the 1920s and contributing over one-third of the total number of published papers in the field by the end of the decade.[3]

A number of formal speeches and more than four hundred scientific papers filled the days of the week-long congress. The program had a decidedly American cast and reflected the scope and diversity of the discipline. Sessions ranged from "Animal Behavior" and "Psychogalvanic Reflex" to "Industrial Psychology" and "Applications of Psychology to Methods of Teaching." The general intellectual tone was behavioristic, set by major addresses given by Ivan Pavlov, the Russian discoverer of the conditioned reflex, and Karl Lashley, the most prominent disciple of John B. Watson.[4] In contrast, psychoanalysis was little in evidence, and Sigmund Freud was not in attendance. Although his ideas were widely known, they had achieved greater acceptance among psychiatrists and the lay public than among professional psychologists.[5]

The International Congress was sponsored by the American Psychological Association (APA), the main professional organization for U.S. psychologists. Begun in 1892, the APA had grown from 127 members at the turn of the century to more than a thousand by 1929. Psychology had become a standard subject in higher education and was among the larger scientific disciplines in America. Although it was a highly academic enterprise, in that a high proportion of its professionals held doctorates and were employed as professors, much of the growth of the field was due to the perceived utility of psychological concepts and techniques,

2 There were 678 counted in the U.S.A. compared to a total of 551 elsewhere. The following leading nations counted for the bulk of the remainder: Germany (142), United Kingdom and British Empire (102), France (90), and the U.S.S.R. (38), with the rest scattered among more than a dozen other countries. Calculated from Carl Murchison, ed., *The Psychological Register* (Worcester, Mass.: Clark University Press, 1929).

3 J. McKeen Cattell, "Psychology in America," in Ninth International Congress of Psychology, *Proceedings and Papers* (Princeton, N.J.: Psychological Review Co., 1930), pp. 12–32, on pp. 26–28; Samuel W. Fernberger, "On the Number of Articles of Psychological Interest Published in the Different Languages (1916–1925)," *American Journal of Psychology,* 1926, *37*:578–581.

4 See Ninth International Congress of Psychology, *Proceedings and Papers.* Zygmunt Pietrowski, "Freudian Causation," pp. 343–344; A. A. Brill, "Psychoanalysis and Conduct Problems," p. 97. Lashley's address, "Basic Neural Mechanisms in Behavior," was subsequently published in the *Psychological Review,* 1930, *37*:1–24.

5 See John C. Burnham, "The New Psychology: From Narcissism to Social Control," in *Paths into American Culture: Psychology, Medicine, and Morals* (Philadelphia: Temple University Press, 1988; orig. pub. 1968), pp. 69–93; Nathan G. Hale, Jr., *Freud in America,* 2 vols. (New York/Oxford: Oxford University Press, 1995); Gail A. Hornstein, "The Return of the Repressed: Psychology's Problematic Relations with Psychoanalysis, 1909–1960," *American Psychologist,* 1992, *47*:254–263.

particularly mental testing. The expansion of the American educational system, both at the secondary level and in higher education, provided many opportunities for psychologists to advance their field.[6]

The officers of the congress were a distinguished group, drawn from the ranks of leading American psychologists, several of whom had played important roles in the establishment of the field. The most senior among them, James McKeen Cattell (1860–1944), served as president. He had been Wilhelm Wundt's first American student forty years earlier and had helped make Columbia University a leading center for research and graduate training at the turn of the century. The role of treasurer was filled by one of Cattell's former students who became his colleague at Columbia, Robert S. Woodworth (1869–1962), then in the middle of an extraordinarily long and productive career as a well-rounded experimentalist. Both Woodworth and the vice president of the congress, James R. Angell (1869–1949), the president of Yale University, had been deeply influenced by their contact with William James as students at Harvard and were vigorous proponents of the eclectic and pluralistic approach known as "functionalism" in psychology. A generation younger, Edwin Boring (1886–1968), head of the Harvard psychology laboratory, served as secretary.

These men, along with fourteen more colleagues, constituted the National Committee for the congress. Representing major graduate programs, they formed the academic elite of a field that had grown enormously in the decade following the First World War. Many of them were veterans of the military psychology programs that had increased the salience and visibility of applied work, both professionally and publicly. That wartime involvement, perhaps more than any other professional experience, provided an enduring point of reference for an entire generation.

The Rise of an Academic Profession

The evolving relationship between scientific research and social applications in the United States was highlighted at the congress by the presidential address delivered by James McKeen Cattell. The sixty-nine-year-old psychologist's speech on "Psychology in America" provided a historical survey of the development of the field. In his view, "the chief contribution of America to psychology has not been large philosophical generalizations, but the gradual accumulation from all sides of facts and methods that will ultimately create a science, both descriptive and applied, of human nature and human behavior."[7] This statement expressed the characteristic faith in Baconian induction held by Cattell and many of his colleagues. Psychology,

6 See Raymond E. Callahan, *Education and the Cult of Efficiency* (Chicago: University of Chicago Press, 1962); John M. O'Donnell, *The Origins of Behaviorism: American Psychology, 1870–1920* (New York: New York University Press, 1985).

7 McKeen Cattell, "Psychology in America," in Ninth International Congress of Psychology, *Proceedings and Papers* (Princeton, N.J.: Psychological Review Co., 1930), pp. 12–32, on p. 21.

because of its fundamental concern with human welfare, demonstrated how the boundaries between descriptive and normative approaches to science had disappeared. Thus, according to Cattell, the aim of psychology was "to describe, to understand and to control human conduct."[8]

Cattell cited quantitative data on the growth of psychology in the United States, noting trends in the membership of the American Psychological Association, publication patterns, and other figures. He observed that eight universities were each spending more than $40,000 a year on research and teaching in psychology, and that fifty-eight others spent at least $10,000 annually. At Columbia, including Teachers College and Barnard College, the budget for psychology totaled nearly $200,000. Cattell's data showed that the bulk of American psychologists were investing their time in working on individual differences, mainly in the context of education.[9]

Although Cattell was a prominent member of the first generation of professional psychologists, his reputation had been built on his successful career in scientific publishing rather than his contributions to research. Among the many publishing ventures he managed was *Science* magazine, which he had revived at the turn of the century and later brought under the sponsorship of the American Association for the Advancement of Science, and the continuing series of *American Men of Science* biographical directories, which had become an indispensable reference tool. Cattell's position allowed him to be a gadfly to the scientific community, and he took delight in opposing what he considered to be elitist control over science by the National Academy of Sciences and the large private foundations that supported research. In his remarks to the International Congress of Psychology, Cattell stressed that progress in psychology depended upon the work of the rank and file as well as the leaders of the discipline.[10]

Cattell's egalitarian beliefs did not prevent him from sharing the podium with one of the prime exponents of the very scientific elitism he opposed. James R. Angell, president of Yale University, was serving as the vice president of the Ninth International Congress of Psychology. In his capacity as local host, he welcomed the participants to America and to Yale in his opening remarks. Angell was another member of the first generation of American psychologists and had had a distinguished career as an educator, first as chairman of the University of Chicago's psychology department for a quarter-century and then as Yale president. He had also held the post of chairman of the National Research Council from 1920 to 1921, when

---

8 Ibid., p. 31.   9 Ibid., p. 26.
10 Michael M. Sokal, ed., *An Education in Psychology: James McKeen Cattell's Journal and Letters from Germany and England, 1880–1888* (Cambridge, Mass.: MIT Press, 1981); idem, "James McKeen Cattell and Mental Anthropometry: Nineteenth-Century Science and Reform and the Origins of Psychological Testing," in Michael M. Sokal, ed., *Psychological Testing and American Society, 1890–1930* (New Brunswick, N.J.: Rutgers University Press, 1987), pp. 21–45.

he oversaw the construction of palatial new headquarters for its parent body, the National Academy of Sciences, in Washington, D.C. Well-connected in foundation circles, Angell had served briefly as president of the Carnegie Corporation, and one of his Chicago doctoral students, Beardsley Ruml, headed the Laura Spelman Rockefeller Memorial.

Intellectually, Angell was a leading exponent of functionalism in psychology and had systematically articulated this approach through a series of influential papers and books.[11] He took an equally pragmatic approach to administration. A vigorous advocate of the research mission of the university, Angell's intellectual and institutional ambitions were inextricably intertwined. His father, James B. Angell, had served for many years as president of the University of Michigan and was among the generation of leaders who helped to transform higher education in America during the late nineteenth century.[12] Cultivated and diplomatic like his father, Angell spent the major portion of his career at the University of Chicago, where the powerful departmental structure provided the institutional context for the pursuit of disciplinary goals. After the First World War, this model of the academic research enterprise was powerfully reinforced by increasing support from government and private foundation sources, and Angell was in the vanguard of those who articulated a progressive vision of research organized and managed for the greater good.[13]

Moving to Yale in 1921, Angell supervised the revival of psychological research, which had lapsed after an auspicious beginning before the turn of the century. With the help of foundation funds, the Institute of Psychology was formed in 1924, and three prominent researchers – Raymond Dodge, Clark Wissler, and Robert M. Yerkes – were hired. These established scientists shared an institutional affiliation but subscribed to no collective programmatic goals beyond the pursuit of their own individual research projects. As a group of independent investigators they provided a nucleus around which even more ambitious plans for an interdisciplinary program would coalesce.[14]

11 See James R. Angell, "The Province of Functional Psychology," *Psychological Review,* 1907, *14*:61–91 (his APA presidential address), or his textbook *Psychology: An Introductory Study of the Structure and Functions of Human Consciousness* (New York: Holt, 1904), which reached a 4th edition by 1908.

12 See Laurence R. Veysey, *The Emergence of the American University* (Chicago: University of Chicago Press, 1965); Burton J. Bledstein, *The Culture of Professionalism* (New York: Norton, 1976).

13 James R. Angell, "The Organization of Research," *Scientific Monthly,* 1920, *11*:26–42. See Roger L. Geiger, *To Advance Knowledge: The Growth of American Research Universities, 1900–1940* (New York: Oxford University Press, 1986).

14 The revitalized psychology program at Yale was already beginning to attract new graduate student interest by the mid-1920s. By the end of the decade seventeen doctorates had been granted, compared to only three during the previous ten years (Robert S. Harper, "Tables of American Doctorates in Psychology," *American Journal of Psychology,* 1949,

In early 1929, several months before the Ninth International Congress of Psychology, Yale announced a major academic initiative. The Rockefeller Foundation pledged nearly $10 million for the establishing of the Institute of Human Relations. The venture drew in the Schools of Medicine and Law and the various social science departments in an attempt to forge an interdisciplinary and cooperative scientific attack on the roots of social and individual problems.[15]

The founding of the Institute of Human Relations reflected the hope that science could provide the basis for a rational management of human affairs. There were great expectations for major advances in the social sciences, as the behavioral paradigm promised to provide a unifying framework. Psychology was at the center of the emerging "kingdom of behavior." Its focus on individual conduct, the role of learning, and strict objective methods proved attractive not only in psychology but in a wide array of other fields.[16]

Institutional Trends

Behaviorism provided some measure of intellectual coherence to psychology during the 1920s as it underwent rapid professional expansion and widespread academic institutionalization. The field shared in the general growth of higher education, and the psychology community grew substantially as new laboratories were created and the discipline achieved independent departmental status in most universities.

In the decade following the war, two dozen new experimental laboratories were created, raising the total to more than a hundred in the country. In 1925 the total capital investment in laboratory equipment for psychology (mostly for experimental work) had risen to more than $1 million, compared to $30,000 in 1893.[17]

62:579–587). Of course, many Yale graduate students attended the International Congress, where they shared the myriad mundane tasks connected with hosting the event. Some, such as Ernest Hilgard (Ph.D. 1930), acted as escorts for the various foreign dignitaries and were thus able to rub shoulders with eminent figures in the discipline.

15 J. G. Morawski, "Organizing Knowledge and Behavior at Yale's Institute of Human Relations," *Isis,* 1986, *77:*219–242. See also James H. Capshew, "The Yale Connection in American Psychology: Philanthropy, War, and the Emergence of an Academic Elite, 1930–1955," in *The Development of the Social Sciences in the United States and Canada: The Role of Philanthropy,* eds. Theresa R. Richardson and Donald Fisher (Greenwich, Conn.: Ablex Publishing Corporation, 1998).

16 See Franz Samelson, "Organizing for the Kingdom of Behavior: Academic Battles and Organizational Policies in the Twenties," *Journal of the History of the Behavioral Sciences,* 1985, *21:*33–47.

17 The U.S.A. had also emerged as a major commercial supplier of psychological apparatus; the leading American manufacturer was selling nearly $500,000 worth of equipment a year worldwide. C. A. Ruckmick, "Development of Laboratory Equipment in Psychology in the United States," *American Journal of Psychology,* 1926, *37:*582–592; C. R.

Such apparatus was used for undergraduate instruction as well as for research, as psychology became omnipresent in college curricula across the country. The number of universities offering the doctorate in psychology also grew in the 1920s, reaching at least thirty-five by the end of the decade.[18]

Although the number of degree-granting institutions expanded during the 1920s as the next wave of laboratory development reached into the lower echelons of higher education, there was a well-defined core of leading centers for graduate education in psychology. Eight universities – Columbia, Chicago, Iowa, Johns Hopkins, Cornell, Stanford, Harvard, and Pennsylvania – accounted for nearly two-thirds of the doctorates produced between 1919 and 1929.[19] Most had developed strong programs in applied psychology, particularly in connection with mental testing and education, which had emerged from the war as a major venue for the practical application of scientific psychology.

Doctorate production rose rapidly in the 1920s, at a rate nearly triple that of the previous decade. By 1929 nearly eighty Ph.D.s a year were being awarded in the United States (see Table 1.1). For the first time substantial numbers of women were earning advanced degrees in psychology. As early as 1921 one out of every five psychologists listed in *American Men of Science* was female. Women faced serious sex discrimination, however, in pursuing graduate education at this time. A few institutions, such as Columbia Teachers College, were havens for aspiring female professionals, whereas others, such as Harvard University, were bastions of male chauvinism.[20]

As the number of professional psychologists increased and the demand for psychological expertise grew, a major shift in employment patterns occurred. Before the war, college and university teaching was the primary occupation of a vast majority of the nation's psychologists; fewer than one out of ten members of the APA held a position in applied psychology in 1916. Ten years later the proportion had doubled. Nearly 20% were employed in nonacademic work, including clinical, school, personnel, industrial, consulting, and research positions.[21]

Changes in the labor market were paralleled by shifts in academic teaching and research. During the 1920s the proportion of APA members who reported applied psychology as a subject of instruction increased nearly threefold, from 11% to 29%,

Garvey, "List of American Psychology Laboratories," *Psychological Bulletin,* 1929, *26:* 652–660. See also James H. Capshew, "Psychologists on Site: A Reconnaissance of the Historiography of the Laboratory," *American Psychologist,* 1992, *47:*132–142.

18 Harper, "Tables of American Doctorates."   19 Ibid.

20 James H. Capshew and Alejandra C. Laszlo, "'We Would Not Take No for an Answer': Women Psychologists and Gender Politics during World War II," *Journal of Social Issues,* 1986, *42*(4):157–180.

21 F. H. Finch and M. E. Odoroff, "Employment Trends in Applied Psychology," *Journal of Consulting Psychology,* 1939, *3:*118–122; idem, "Employment Trends in Applied Psychology, II," *Journal of Consulting Psychology,* 1941, *5:*275–278.

Table 1.1. *U.S. doctorate
production in psychology,
1900–1959*

Years	Ph.D.s awarded
1900–1904	57
1905–1909	85
1910–1914	111
1915–1919	119
1920–1924	217
1925–1929	427
1930–1934	545
1935–1939	571
1940–1944	528
1945–1949	725
1950–1954	2,752
1955–1959	3,650

Sources: Robert S. Harper, "Tables
of American Doctorates in Psychology," *American Journal of Psychology,* 1949, *62:*579–587; Lindsey R.
Harmon, *A Century of Doctorates*
(Washington, D.C.: National Academy of Sciences, 1978), p. 13.

while the percentage teaching experimental psychology courses declined slightly.
Research areas exhibited similar changes. Those reporting applied research interests rose from 37% to 48% over the same period, whereas those reporting interest
in experimental psychology declined.[22]

Increasing doctorate production and an expanding employment base contributed
to the growth of the APA after the war. By 1930 it had nearly tripled in size, reaching a total of 1,100 members. Due in part to the wartime success of mental testing,
several prominent applied psychologists were elected to the presidency of the APA,
including Walter Dill Scott (1919), S. I. Franz (1920), and Lewis M. Terman (1923).

Intellectual Diversity

The professional growth of psychology during the 1920s was accompanied by intellectual ferment. Behaviorism was on the rise, challenging earlier formulations

22 Samuel W. Fernberger, "Statistical Analyses of the Members and Associates of the
American Psychological Association, Inc. in 1928: A Cross Section of American Professional Psychology," *Psychological Review,* 1928, *35:*447–465.

that stressed psychology as the science of mind or consciousness. Other points of view, such as Gestalt psychology, also vied for attention. Popular interest in psychology reached faddish levels as well. One commentator noted the "outbreak of psychology" that seemed to be sweeping the country.[23]

The 1920s have often been characterized as the period of the "schools" in American psychology, when a variety of theories and systems competed with each other for intellectual dominance. Different schools were associated with different universities, so it was possible to map the cognitive structure of the field onto its institutional structure, at least in a general way. Of course, the sociointellectual ecology was dynamic and diverse, and the conceptualization of psychology in terms of competing schools of thought can be seen as a way to manage the heterogeneity of the field's aims and methods.

The use of schools of psychology as an organizing principle can also be seen as an attempt to address the perennial problems involved in educating the next generation of scientists. Schools represented not only theories but also practices, and they embodied social as well as epistemological values. In the process of being trained in the craft of science, aspiring psychologists were imbued with the ideological commitments of their mentors. As students were socialized into the profession in various academic programs, they were exposed to the characteristic intellectual prejudices and predilections of their departments. Thus different universities became noted for particular approaches, such as Cornell's structuralism, Chicago's functionalism, and Johns Hopkins's behaviorism. Naturally, these identities changed over time and were often associated with a single major figure, such as Titchener at Cornell or Watson at Hopkins.

In an expanding market for psychological knowledge and professional expertise, the lack of strong theoretical consensus among psychologists was viewed as a sign of healthy disagreement. Nonetheless, it did represent a potential threat to disciplinary unity and professional harmony. As the psychology community grew, discussions of the merits and defects of different points of view or schools of thought became a staple of professional discourse. A market emerged for books that dealt with such issues. Aimed at aspiring professionals as well as members of the interested public, such volumes displayed the theoretical wares of psychology and, in some cases, attempted to guide the consumer in purchasing such intellectual products.

Psychologies of 1925 (1926) was paradigmatic of this genre. As its title suggests, it expressed the particularism of contemporary psychological thought within an implied framework of pluralism. It contained major sections on behaviorism, dynamic psychology, Gestalt psychology, purposive psychology, reaction psychology, and structural psychology. No attempt was made to summarize or provide a systematic

23 Stephen Leacock, "A Manual of the New Mentality," *Harper's,* 1924, *148:*471–480. See also Burnham, "The New Psychology"; Sokal, "James McKeen Cattell and American Psychology in the 1920s."

critique of the competing claims of each psychology, although sharp theoretical disagreements were often expressed by individual authors.

The volume was justified by its organizer, Carl Murchison of Clark University, on pedagogical grounds. He thought it was important to present the diversity of schools of psychology without lapsing into caricature. He noted that "theoretical tradition becomes established in certain educational communities, and students are born structuralists or behaviorists just as one may be born a democrat or a presbyterian."[24] Hence it was important for students to learn about traditions other than their own, lest sectarian prejudice overwhelm the urge for common intellectual endeavor.

Another approach to managing disciplinary diversity in the name of scientific progress was taken by authors of historical treatises on psychology. The ambitions of American psychologists for the future were built on an increasing awareness and appreciation of their collective past, and by the late 1920s a self-conscious and sustained interest in the history of the psychology had emerged. History, with its developmental logic, offered a way to integrate an epistemologically unruly field and an alternate means of defining a disciplinary identity. Nineteen twenty-nine proved to be a banner year for psychologist-historians, with the appearance of three major books on the history of the field, each of which conveyed a different message about its possible future.

Closest in spirit and approach to *Psychologies of 1925,* Walter Pillsbury's *History of Psychology* (1929) was motivated by similar pedagogical concerns. The aim of the book was to provide "a historical setting to current controversies" and to suggest the relevance of past ideas to present concerns.[25] A straightforward and descriptive textbook, it was apparently directed toward an undergraduate audience. Pillsbury, a well-known textbook author, was the longtime chairman of the University of Michigan psychology department and a former student of Titchener at Cornell. His history traced the philosophical background of modern psychology to the emergence of experimental psychology in the nineteenth century. Later chapters dealt with developments in the national contexts of Germany, France, and the United States, and with psychoanalysis, structuralism, functionalism, behaviorism, and hormic and Gestalt psychology. The survey concluded with a brief remark on the variety of current schools in the field, suggesting that intellectual progress would be made through some process akin to natural selection. Content with explicating the status quo, Pillsbury's book straddled the fence on the issue of whether psychology should be considered the science of mind or of behavior.

The second volume in this accidental trio of histories was *An Historical Introduction to Modern Psychology* by Gardner Murphy, a young instructor in psy-

24 Carl Murchison, "Preface," in *Psychologies of 1925* (Worcester, Mass.: Clark University, 1926), p. v.
25 W. B. Pillsbury, *The History of Psychology* (New York: W. W. Norton, 1929), p. 10.

chology at Columbia University. His attempt to put contemporary psychology in perspective was based on the idea that the rise of the new psychology in the nineteenth century derived from "the interaction of experimental physiology, psychiatry, the theory of evolution, and the social sciences, constantly working upon materials from the history of philosophy, and guided by progress in the physical sciences and statistical method."[26] Murphy's broad-gauged approach yielded a book that placed the development of psychology in the context of the rise of science since the seventeenth century, and he included recent developments, including psychoanalysis, child psychology, social psychology, and personality. In his conclusion, he stressed the importance of genetic (i.e., developmental) and statistical methods that had come to augment traditional experimentation. Indeed, he noted that quantification in general and mathematization in particular had become the sine qua non of modern science. He speculated that this would lead to a convergence of psychology and physiology, but wondered whether certain aspects of the self and personal experience could be adequately described within a framework of natural science.

Murphy's book portrayed psychology as a pluralistic discipline, embedded in a complex web of interrelationships with other branches of knowledge, the education system, and the practice of the healing arts. By appealing to the rich origins of the field, it provided a rationale for a diversity of views and eclecticism in method. It also pointed to personality as an emerging focus for research and practical applications.

The third historical treatise to appear in 1929 was *A History of Experimental Psychology* by Edwin Boring. More scholarly than Pillsbury's text and more narrowly focused than Murphy's survey, Boring's opus appropriated the figure of Wilhelm Wundt as the founder of modern psychology as an experimental science. Like Pillsbury, Boring had received his doctorate at Cornell under Titchener, but he had taken on the views of his mentor much more self-consciously and completely than his colleague. Outwardly, Boring's *History* was a vigorous defense of traditional experimental psychology; it was also an implicit declaration of the independence of psychology from philosophy. Drawing a distinct boundary between philosophy and psychology was important to Boring for more than purely intellectual reasons. At Harvard, he had been struggling for years to create a department of psychology separate from philosophy – a battle that had already been won at practically all other major universities. Not until 1932 did Boring succeed.

Boring's history was also notable because in it he assumed the mantle of his mentor Titchener (who died in 1927) as the scourge of the discipline's scientific conscience. As the field moved in new directions, it seemed more important than ever to reaffirm a fundamental faith in experimental research in order to legitimate

---

26 Gardner Murphy, *An Historical Introduction to Modern Psychology* (New York: Harcourt, Brace, 1929), p. xvi.

various kinds of social applications and personal interventions. Endorsing Boring's historical thesis, if only through silent acquiescence, provided a way in which contemporary practitioners of all varieties could display their scientific piety without infringing on their existing disciplinary and professional commitments. In other words, Boring's history served a useful function for the entire profession by providing a scientific pedigree that was both impressive and plausible, to insiders and outsiders alike. Even if his fundamental contention that the development of psychology depended upon the pursuit of scientific knowledge for its own sake was debatable, the erudite product of his historical labors played an enduring role in the self-image of the profession.[27]

In distinctly different ways, each of these three volumes expressed the faith that the past was not only intelligible but relevant to the present. Moreover, they all concentrated on intellectual developments – theories, methods, findings – rather than on the application of psychological knowledge or the community of psychologists. Murphy saw psychology as a product of social and cultural forces as well as scientific and technical developments. Although Pillsbury's historiographical assumptions are more obscure, apparently he believed that history could be useful in helping to guide students through the perplexing proliferation of contemporary schools of thought. Boring, with his personalistic emphasis on the "great men" of psychology, attempted to reassert the centrality of experimentation in an increasingly heterogeneous enterprise.

By the late 1920s experimentalists in the discipline were becoming increasingly concerned about the shift away from traditional laboratory research toward applied work and made efforts to stem the rising tide. In this context, Boring's *History of Experimental Psychology* can be viewed as an attempt to assert the primacy of laboratory methods and findings in the establishment of the discipline. By identifying the roots of modern psychology with the application of physiological techniques to the study of sensation, perception, consciousness, and other areas previously subsumed under philosophy, Boring was staking a claim that the future of psychology lay in the progressive elaboration of this research tradition, begun by Fechner and continued by Wundt, Titchener, and others. The fact that Boring felt compelled to write such a book suggests how seriously he perceived a threat to the traditional source of cognitive authority in the field. Despite his puritanical scientism, he was not completely unappreciative of mental testing and other forms of psychological investigation. Nonetheless, he was a leader in the movement to restore experimentalism as the unquestioned essence of psychology.[28]

27 See Mitchell G. Ash, "The Self-Presentation of a Discipline: History of Psychology in the United States between Pedagogy and Scholarship," in *Functions and Uses of Disciplinary Histories,* ed. L. Graham, W. Lepenies, and P. Weingart (Dordrecht: Reidel, 1983), pp. 143–189.

28 John M. O'Donnell, "The Crisis of Experimentalism in the 1920s: E. G. Boring and His Uses of History," *American Psychologist,* 1979, *34:*289–295. Cf. Franz Samelson,

Some of Boring's sympathetic colleagues recognized the propaganda value of his history. Leonard Carmichael, director of the psychological laboratory at Brown, called the study "an event of the first magnitude." He considered it a manifesto for the cause:

> It is difficult to imagine that a psychologist can read this book and still not feel the lure of a new working plan for his science. Such a plan would emphasize experiment as never before. . . . This development, however, must be in the hands of scientists who are almost harshly empirical and who, like Dr. Boring himself, yield to no emotional desire to leave the laboratory in order to save society.[29]

With assertions about the values and attitudes of psychologists, the battle between experimentalism and its challengers was thus joined on moral as well as episto mological grounds.

Another key defender of the experimentalist faith was Knight Dunlap, head of the Johns Hopkins psychology department. In a 1929 essay, he surveyed "The Outlook for Psychology" and found the prospect not entirely reassuring. He believed that the academic base of psychology had been eroded by the rise of applied work, and that steps should be taken to strengthen college and university departments of psychology. More research and less teaching was his prescription; he urged psychologists to get back to their laboratory benches and do more experiments.[30]

Dunlap was also active on the organizational front. In late September 1929 he helped create the National Institute of Psychology in Washington, D.C. Established as a national research center for "problems in human and animal psychology," the new enterprise was incorporated by Dunlap, Hugh S. Cumming, the U.S. surgeon general, and Edwin E. Slosson, director of the Science Service news organization. It was envisioned as a central laboratory where university-based researchers could spend summers and other periods of leave performing investigations "too long, expensive, and complicated for other institutions." The use of the "cooperative method" would increase the quality and efficiency of the research. Its function would be similar to the National Bureau of Standards, the federal government's central physical laboratory, but it would be privately operated. The announcement listed fifty prominent experimental psychologists as charter members but contained no details on financial or managerial arrangements.[31] Plans for the National Institute

"E. G. Boring and His *History of Experimental Psychology,*" *American Psychologist,* 1980, *35:*467–470.

29 Leonard Carmichael, review of *A History of Experimental Psychology, New Republic,* 5 November 1930, *64:*330–331, quote on p. 331.

30 Knight Dunlap, "The Outlook for Psychology," *Science,* 1929, *69:*201–207.

31 "To Study Problems in Psychology," *New York Times,* 28 September 1929, p. 40. The members listed were: John Anderson (Minnesota), Madison Bentley (Cornell), E. G. Boring (Harvard), Warner Brown (California), Harvey Carr (Chicago), Percy Cobb (Cleveland), J. E. Coover (Stanford), S. [*sic*] K. Dallenbach, (Stanford [*sic*] Cornell),

of Psychology never came to fruition, even though the organization continued to exist, on paper at least, through the 1930s.

The Aftermath of the Depression

No doubt part of the reason the National Institute of Psychology never got off the ground was that the U.S. stock market crashed within weeks of the announcement of its formation. After a decade of relative economic prosperity following the Great War, the Great Depression had begun. It affected nearly every facet of American life, including higher education, and psychology was not exempt from the stagnant employment market it caused or the social problems it generated.

The supply of professional psychologists continued to grow through the 1930s and what little expansion occurred in the labor market took place mainly in nonacademic positions. Retrenchment in the university was the order of the day, and the shift away from experimental toward applied psychology continued. As competition for resources increased, intellectual debates over the aims and practice of psychology intensified, and newly organized interest groups arose within the field to challenge traditional power structures. At the end of the 1930s, one observer noted, psychology was facing its "impending dismemberment" along ideological, organizational, and generational lines.[32]

Ironically, many of the problems the psychology community faced in adjusting to the depression derived from its success in establishing a strong academic presence following World War I. As psychology was institutionalized in American universities, it developed a self-sustaining culture of research, training, and practice. In an expanding system of higher education, new Ph.D.s could be easily absorbed into jobs that replicated the positions of their professors. But when demand dropped, it proved difficult to decelerate the enormously productive Ph.D. machine. After large increases during the 1920s, doctorate production doubled in the 1930s, when an average of 100 new Ph.D.s were conferred each year.

This flood of newly credentialed recruits had several important consequences for the profession. It made chronic employment problems even more acute and in-

J. S. Dashiell (North Carolina), F. C. Dockeray (Ohio Wesleyan), R. Dodge (Yale), K. Dunlap (Hopkins), F. Fearing (Northwestern), S. W. Fernberger (Pennsylvania), Frank Freeman (Chicago), L. Hollingworth (Columbia), C. Hull (Wisconsin), W. Miles (Stanford), H. S. Langfeld (Princeton), Joseph Peterson (Nashville), R. Pintner (Columbia), A. T. Poffenberger (Columbia), E. S. Robinson (Yale), C. P. Stone (Stanford), M. F. Washburn (Vassar), A. P. Weiss (Ohio State), H. Woodrow (Oklahoma), R. S. Woodworth (Columbia), R. Yerkes (Yale), H. M. Johnson (Mellon Institute), K. S. Lashley (Institute for Juvenile Research), F. L. Wells (Boston Psychopathic Hospital).

32 Heinrich Klüver, "Psychology at the Beginning of World War II: Meditations on the Impending Dismemberment of Psychology Written in 1942," *Journal of Psychology,* 1949, 28:383–410.

creased the resort to nonacademic labor markets. It also led to a significant change in the age structure of the profession, so that by 1939 approximately half of all Ph.D. psychologists had had their doctorates for less than ten years.[33] It contributed to the increasing ideological and organizational fragmentation of the field.

Many members of this depression-era generation went into applied areas. Some entered by choice; others went more reluctantly in response to financial exigencies. Employment trends in the American Psychological Association showed a strong increase in applied work in the decades after the war. In 1916 the vast majority of APA members were college and university teachers, with less than one out of ten holding applied positions. By the middle of the 1930s more than one in three were employed in applied psychology, including clinical, school, personnel, industrial, consulting, and research positions.[34]

Statistics are unavailable on the number of psychologists who left the field during the depression, but their numbers may have been considerable. In response to concerns over the declining job market, in 1933 the APA conducted a study of the supply and demand for psychologists. The findings were sobering. In 1931/1932 there were 100 new doctoral recipients but only 33 new positions available. The outlook for those with master's degrees was even more bleak: there were 405 new degree holders competing for 40 new jobs. And there was little hope that the situation would improve over the next several years.[35]

Despite the expansion of applied psychology, however, graduate training continued to emphasize experimental psychology. In an attempt at quality control, the APA appointed a committee on the Ph.D. degree that conducted a fact-finding survey of major graduate programs. The results, published in 1934, revealed variation in admissions criteria and formal degree requirements but a strong consensus that training in experimental psychology was fundamental, even for those going into abnormal, child, and social psychology. Nearly all the programs required course work in experimental psychology, statistics, and history and theory, as well as reading knowledge of German and French.[36] These findings suggested that the pedagogic core of the discipline was retaining its scientific emphasis in the face of changing professional interests and employment patterns.

Heavy doctorate production and an expanding occupational base contributed to the growth of the American Psychological Association in the interwar period. Between 1915 and 1945 membership multiplied approximately tenfold, from 300 to

33 Stephen Habbe, "A Comparison of the American Psychological Association Memberships of 1929 and 1939," *Psychological Record*, 1940–1941, *4*:215–232.

34 Finch and Odoroff, "Employment Trends in Applied Psychology."

35 A. T. Poffenberger et al., "Report on the Supply and Demand for Psychologists Presented by the Committee on the Ph.D. in Psychology," *Psychological Bulletin*, 1933, *30*:648–654.

36 APA Committee on the Ph.D. Degree in Psychology, "Standards for the Ph.D. Degree in Psychology," *Psychological Bulletin*, 1934, *31*:67–72.

Table 1.2. *American Psychological Association membership, 1900–1985*

Year	Totals (incl. Fellows/Members/Associates)	(Associates)[a]
1900	127	
1905	162	
1910	228	
1915	295	
1920	393	
1925	471	
1930	1,101	(571)
1935	1,818	(1,276)
1940	2,739	(2,075)
1945	4,173	(3,161)
1950	7,273	(5,775)
1955	13,475	(11,579)
1960	18,215	(14,008)
1965	23,561	(16,664)
1970	30,839	(21,502)
1975	39,411	(28,552)
1980	50,933	(38,675)
1985	60,131	(47,901)

[a]Changed to "Members" after 1957.
Sources: Robert S. Daniel and C. M. Louttit, *Professional Problems in Psychology* (New York: Prentice-Hall, 1953), p. 17; American Psychological Association Library.

3,000. Most of the growth, however, was confined to the subordinate "associate" membership class. Carrying no voting privileges, this category was created in 1926 by conservative APA leaders who were concerned about the drift away from the laboratory and wanted to blunt the influence of applied psychology. This effectively marginalized younger psychologists working in nonacademic positions, since full APA membership was conferred only upon those having a substantial publication record and with the approval of the senior members. This tactic preserved the organizational power of an older generation of experimentalists at the expense of marginalizing newly emerging groups (Table 1.2).[37]

Generational strains were complicated by a growing gender gap. During the 1920s and 1930s increasing numbers of women were making careers in psychology. One-fifth of the psychologists listed in the 1938 edition of *American Men of Science* were women. Women accounted for an even larger proportion – nearly

37 On the creation of the APA Associate category, see O'Donnell, "The Crisis of Experimentalism in the 1920s."

one-third – of the membership in the two major professional societies, the American Psychological Association and the American Association for Applied Psychology (AAAP). By the beginning of World War II, it was estimated that well over 1,000 women held graduate degrees in psychology, including more than 25% of the doctorates.[38]

Although psychology was a highly feminized discipline, trailing only nutrition in its percentage and zoology in its absolute numbers of female scientists, few women managed to attain leadership positions within professional organizations. For instance, only two women had been elected president of the APA since its founding in 1892 – Mary Calkins in 1905 and Margaret Washburn in 1921.[39] These accomplished women had achieved eminence when the APA was still a small, comparatively intimate group with fewer than 400 members; as the organization grew to nearly 3,000 members by 1940, conservative nominating committees effectively limited the chances for other women to hold office. By comparison, no woman had been elected president of the AAAP, formed in 1937, and only two served on its 29-member founding board. Women were, however, better represented in the AAAP as division chairs, journal editors, and administrative officers than in the APA, where they held virtually no offices.[40]

The exclusionary practices of the APA extended to rank-and-file members as well. Between 1923 and 1938 the APA had experienced an extraordinary rise in female membership: from 81 women in 1923 (18% of the total) the number multiplied eightfold to 687 (or 30%) in 1938. Perhaps alarmed by this unprecedented influx, in 1925 several of the APA's elite, prompted by Harvard psychologist Edwin G. Boring, suggested establishing a two-tier membership with a distinguished category for fellows (originally termed "members"). Although a few of the leaders polled were concerned that a differentiated membership might give rise to professional jealousies and make the society "undemocratic," the majority approved the measure as a means of quality control. The associate category was subsequently introduced in 1926 with little opposition; all current APA members would become fellows, but new associates who wanted to gain such status would have to be elected by the usually all-male council. Although no one at the time expressed concern that the new system might affect men and women differently, by the early 1940s women psychologists finally recognized that it had operated to restrict their

38 Donald G. Marquis, "The Mobilization of Psychologists for War Service," *Psychological Bulletin,* 1944, *41:*469–473; Margaret W. Rossiter, *Women Scientists in America: Struggles and Strategies to 1940* (Baltimore: Johns Hopkins University Press, 1982).

39 On the careers of Calkins and Washburn, see Elizabeth Scarborough and Laurel Furumoto, *Untold Lives: The First Generation of American Women Psychologists* (New York: Columbia University Press, 1987).

40 L. J. Finison and L. Furumoto, "Status of Women in American Psychology, 1890–1940, Or on How to Win the Battles Yet Lose the War," unpublished paper, 1980; Mildred B. Mitchell, "Status of Women in the American Psychological Association," *American Psychologist,* 1951, *6:*193–201; Rossiter, *Women Scientists in America.*

position within the APA. The statistical evidence was indisputable: while the percentage of women in the society had risen from 18% to 30% between 1923 and 1938, the number of female fellows rose only 1%, from 18% to 19%.[41]

The prevalent sex discrimination encountered by women psychologists was also reflected in the dual labor market that had evolved in psychology. Male Ph.D.s tended to hold higher-status jobs in university and college departments, concentrating on teaching and research. Female Ph.D.s, on the other hand, were usually tracked into service-oriented positions in hospitals, clinics, courts, and schools. Discouraged and frequently prevented from pursuing academic careers, women filled the ranks of applied psychology's low-paid, low-status workers. The few women who managed to gain academic employment were mostly relegated to women's colleges, and to university clinics and child welfare institutes linked to departments of psychology and education.[42]

Ideological Disarray

The changing demography of the profession was accompanied by deepening ideological divisions. The contrast between pure and applied psychology provided a fundamental demarcation, but debates over the definition and role of psychology ranged widely across a myriad of philosophical and professional distinctions. Observers within and without noted a sense of intellectual crisis in the psychology community as it struggled to find some common ground among a constituency that was growing not only larger but more diverse.

The changed tone of the debate over the definition of psychology can be gauged by differences between *Psychologies of 1925* and its successor, *Psychologies of 1930*.[43] Although they both employed the same method of simply juxtaposing different points of view, the existence of such diversity was now viewed in a negative light. The contents of *Psychologies of 1930* were even more disparate than those of its predecessor, leading one commentator to suggest that it "revealed such a heterogeneity of aims, methods, and principles that all efforts to arrive at a satisfactory delineation of the field seemed futile."[44]

41 Mitchell, "Status of Women"; O'Donnell, "The Crisis of Experimentalism in the 1920s"; Rossiter, *Women Scientists in America.*

42 James Capshew, Richard Gillespie, and Jack Pressman, "The American Psychological Profession in 1929: Preliminary Data and Results of a Prosopographical Study," unpublished paper, 1983; Samuel W. Fernberger, "Academic Psychology as a Career for Women," *Psychological Bulletin,* 1939, *36:*390–394; Finison and Furumoto, "Status of Women"; Gladys D. Frith, "Psychology as a Profession," *Women's Work and Education,* 1939, *10*(3):1–3.

43 Carl Murchison, ed., *Psychologies of 1925* (Worcester, Mass.: Clark University Press, 1926); Murchison, ed., *Psychologies of 1930* (Worcester, Mass.: Clark University Press, 1930).

44 Klüver, "Psychology at the Beginning of World War II," p. 387.

The prolonged economic depression of the 1930s made psychology's precarious academic identity even more problematic. Financial exigencies forced some universities to retrench and scale back their programs. Lean times also made intellectual controversies sharper, as proponents of differing views fought over limited resources. These conditions led to much soul-searching among American psychologists for ways to deal with a pervasive sense of crisis.[45]

During the depression writing on the past and present state of psychology became even more preoccupied with contemporary debates over competing systems and schools of thought. The welter of worldviews represented by behaviorism, psychoanalysis, functionalism, Gestalt, and a host of other theories, approaches, and perspectives defied easy categorization or summary. At Columbia University, Robert Woodworth wrote a textbook on *Contemporary Schools of Psychology,* published in 1931. Based on his undergraduate survey course, the book covered introspective psychology, behaviorism, Gestalt psychology, psychoanalysis, and purposivism. Woodworth tended to be descriptive rather than critical in his review and saw the proliferation of schools as a natural and healthy sign of scientific growth. For him, "each school began as a revolt against the established order," and progress occurred as once revolutionary movements became established and then, in turn, were overthrown. In the final chapter, "The Middle of the Road," Woodworth revealed the place where he – and probably most other psychologists   felt most comfortable. He advocated a pragmatic eclecticism for the majority of psychologists who were reluctant to ally themselves completely with any single doctrine or point of view. Woodworth also found disciplinary solidarity in theoretical diversity – indeed, much more than an outsider "might expect from the loud noise that has reached his ears."[46]

The loud noise Woodworth referred to was generated in part by the depiction of psychology in the mass media. The promises of scientific psychology were a favorite topic for journalists, and psychologists themselves played a large role in promoting themselves and their work.[47] Among the most thoughtful and astute observers of the

45 The "crisis" in psychology was international in scope. Both Karl Bühler in Germany, in *Die Krise der Psychologie* (Jena: Fischer, 1927), and Lev Vygotsky in the U.S.S.R., in an unpublished monograph on "The Historical Meaning of the Crisis of Psychology," completed in 1926–1927, discussed the proliferation of schools as evidence that the field was in crisis; they differed on its implications for the future of the discipline. See Csaba Pléh, "Two Conceptions on the Crisis of Psychology: Vygotsky and Bühler," in Achim Eschbach, ed., *Karl Bühler's Theory of Language* (Amsterdam/Philadelphia: John Benjamins, 1988), pp. 407–413. On Vygotsky's analysis, see Alex Kozulin, *Psychology in Utopia: Toward a Social History of Soviet Psychology* (Cambridge, Mass.: MIT Press, 1984), pp. 113–116; and David Joravsky, *Russian Psychology: A Critical History* (Oxford: Basil Blackwell, 1989), pp. 262–266.

46 Robert S. Woodworth, *Contemporary Schools of Psychology* (New York: Ronald, 1931), pp. 4 and 217.

47 See John C. Burnham, *How Superstition Won and Science Lost: Popularizing Science*

psychological scene was Grace Adams, a writer who had earned her Ph.D. at Cornell in 1922 under Titchener. After teaching at Goucher College for a short time, she turned to journalism. Possessing the specialized knowledge of an insider and the critical distance of an outsider, she understood that "interest in mental phenomena is alive today not because psychology holds out any great promise of immediately unifying its activities, but just because its present trends are so diverse."[48]

In 1931 Adams published a trade book with the provocative title *Psychology: Science or Superstition?* After giving a lively and informed sampling of the various "brands" of psychology available in America, including all the major schools of thought, Adams asked in her last chapter, "but is it science?" Her answer was negative, "for science is a discipline of general and impersonal facts, and psychology remains a collection of personal and antagonistic theories."[49] For Adams, the more interesting question was:

> Do psychologists want to be scientists? All of their protestations, all their laboratories, their tedious researches, their technical treaties [*sic*] seem to answer that they do. Science is a magic word today, and every single psychologist, no matter what his private beliefs, has profited enormously through the circumstance that Wilhelm Wundt proclaimed, and to a certain extent proved, that psychology could be as exact as any natural science.[50]

She went on to argue that, insofar as psychology could be scientific, it would have to accept the judgment of modern biology expressed by Thomas Henry Huxley, namely, that humans are conscious automata. But most psychologists, with exceptions like Titchener and Watson, were reluctant to accept this judgment and sought ways to balance the deterministic view of modern science with the traditional belief in individual agency and free will. Most psychologists, Adams felt, were comfortable in this realm of metaphysics, as their prolific theorizing demonstrated.[51]

Some American psychologists may have recognized the intellectual predicament described by Adams. But many probably found the subtle message of scientific progress conveyed by Edna Heidbredder's *Seven Psychologies* (1933) more appealing. This volume blended the genres of systematic exposition and historical review more successfully than other works. Well-organized and gracefully written, the book appeared at a time when the novelty of clashing points of view was wearing off and a theoretical stocktaking seemed in order. The seven psychologies selected by Heidbredder were: Titchener and structuralism, the psychology of

and Health in the United States (New Brunswick, N.J.: Rutgers University Press, 1987); Sokal, "James McKeen Cattell and American Psychology in the 1920s."

48 Grace Adams, *Psychology: Science or Superstition?* (New York: Covici-Friede, 1931), p. 8.

49 Ibid., p. 274.   50 Ibid.   51 Ibid., pp. 275–282.

William James, functionalism and the University of Chicago, behaviorism, dynamic psychology and Columbia University, Gestalt psychology, and Freud and the psychoanalytic movement. These rubrics convey the heterogeneous nature of these schools or systems: ideological markers (i.e., various "isms"), personal names, and institutional contexts were used singly or in combination to identify major intellectual positions.

Seven Psychologies was optimistic in tone. Heidbredder, a psychologist at Wellesley College, recognized that lack of unity could be viewed as a serious problem for a scientific field. In countering that criticism, she refused to take the easy way out by searching for some common denominator of the different systems or by arguing that the rise and fall of schools expressed some evolutionary pattern of development. Instead she emphasized the relative youth of the field, pointing out that it was a "science that had not yet made its great discovery," and thus by implication had not yet reached maturity.[52] And the way to scientific discovery was through the use of scientific method for the identification of facts. As knowledge accumulated, in good Baconian fashion, theoretical frameworks would become irrelevant. Once the destination of scientific success had been reached, the various schools and systems that had served as vehicles could be abandoned.[53]

Woodworth, Heidbredder, and others preached the virtues of eclecticism, which found broad appeal among the many psychologists unable or unwilling to subscribe to a particular theoretical dogma. In 1930 Boring noted that "eclectics probably constitute the majority of psychologists," who were either unconcerned about arriving at a single definition of psychology or derived one in an inductive, post hoc fashion based on the actual activities of psychologists.[54] Theoretical diversity and conflict, however, did not prevent psychologists from displaying an impressive professional "solidarity," as Woodworth had noted. This professional solidarity was rooted in the psychology community's widespread aspirations toward scientific legitimacy and reinforced by long-standing traditions of academic acculturation.

Thus, whatever distress psychologists felt over the lack of intellectual consensus within their discipline was mitigated, at least in part, by their shared professional culture. But this shared professional culture depended in important ways on the availability of institutional resources for the production of certified psychological knowledge and the reproduction of credentialed psychologists. The deepening depression made such resources, both financial and moral, more difficult to obtain and led to growing tensions in the profession.[55]

---

52 Edna Heidbredder, *Seven Psychologies* (New York: Century, 1933), p. 425.
53 Ibid., pp. 428–429.
54 Boring, in *Psychologies of 1930;* quoted in Klüver, "Psychology at the Beginning of World War II," p. 392.
55 The Yale Institute of Human Relations represented an important exception to the general pattern of academic retrenchment in psychology. Even though the grand scheme

Organizational Fragmentation

The ideological disarray of psychology in the 1930s was accompanied by the pro-liferation of special-interest groups. Fed by the continuing growth and increasing diversity of the profession, a host of new voluntary associations arose as psychologists organized themselves to pursue a variety of academic and social agendas. The American Psychological Association remained a dominant force, but found its hegemony in professional affairs challenged by new groups.

The small, close-knit professional networks that conservative APA leaders wished to preserve were giving way to larger and more heterogeneous groupings during the interwar years. New centers of training and research had emerged, publication outlets had multiplied, and a host of new organizations were formed to serve a variety of interest groups. The professional world in which psychologists moved was becoming more complex. Psychologists located their professional identities along multiple and interrelated dimensions, including disciplinary specialty, occupational roles, political orientation, methodological preferences, and philosophical assumptions. An already diverse field was becoming even more diverse.

Among the consequences of this situation was the rise of new professional associations that challenged the hegemony of the APA by providing alternative venues for the pursuit of professional goals. The major rival to the APA was the American Association for Applied Psychology (AAAP), formed in 1937. Its aim was not to displace the APA but to offer a forum for those interested in the professionalization of applied psychology. Membership requirements were even stiffer than the APA's: in addition to the Ph.D., prospective candidates for fellow status had to have at least four years of professional experience or substantial published research. The major applied specialties were represented by sections for industrial, clinical, educational, and consulting psychology. In 1940 AAAP membership stood at over 600, nearly equal to the number of full members of the APA. But nearly 90% of AAAP members also belonged to the APA. And despite its applied orientation, the AAAP drew the bulk of its presidential leadership from the ranks of academics.[56]

---

upon which it had been founded was never fully realized, it had been launched with such ample resources that it could carry on in the face of worsening economic and institutional conditions in professional psychology. In many respects it functioned as an oasis in the academic desert of the depression. The new culture of inquiry in psychology that was emerging was perhaps nowhere more in evidence than at the Yale Institute of Human Relations. This experiment in managing the production of useful knowledge proved to be a harbinger of things to come in the world of academic psychology. See Capshew, "The Yale Connection in American Psychology."

56 Of the AAAP's eight presidents (1937–1944), only two had spent major portions of their careers outside of academe.

Other groups sprang up in the 1930s as psychologists reacted against the APA's exclusive concern with traditional scientific issues. The Society for the Psychological Study of Social Issues (SPSSI) and the Psychologists' League were both formed in response to the depression crisis and tried to engage psychology more directly with social and political change, including the creation of more jobs for unemployed psychologists.[57] Other groups coalesced around technical issues. For instance, psychologists interested in exploring the scientific basis of psychological testing and measurement started the Psychometric Society in 1934, which fostered relations between the creators and users of such techniques.[58]

The APA's leadership, which had resisted efforts to move the organization beyond its narrowly defined role as a learned society, was increasingly on the defensive as the new groups proliferated. However, it still commanded the allegiance of a majority of psychologists, who did not wish to repudiate their scientific heritage.[59]

Conclusion

The convergence of occupational, epistemological, and generational strains in the 1930s is exemplified by the case of George A. Kelly (1905–1966). After a varied program of studies that included sociology, labor relations, speech pathology, and anthropology, Kelly earned a Ph.D. in psychology at the University of Iowa in 1931 at the age of twenty-six. Considering himself lucky to land any academic job during the depression, he became a faculty member at Fort Hays State College in his native Kansas, an institution that served the rural population of the western half of the state. As a member of a small psychology department with a practical mission, Kelly decided "to pursue something more humanitarian than physiological psychology."[60]

Among his tasks was the practice and teaching of clinical psychology, including the operation of a traveling clinic that served the scattered communities of western

57 Lorenz J. Finison, "The Psychological Insurgency: 1936–1945," *Journal of Social Issues,* 1986, *42*(1):21–33.

58 Jack W. Dunlap, "The Psychometric Society – Roots and Powers," *Psychometrika,* 1942, *7*:1–8.

59 As academic purists and applied professionalizers fought their ideological battles with each other, they ignored a potential problem that could damage them both: the indeterminate number of credentialed psychologists who eschewed affiliation with any national psychological organization. Based on available evidence, however, there is little indication that this was a significant problem. Applied psychology groups, in particular, had to contend with a certain amount of disaffiliation within their ranks, given the employment of many of their members in positions that rewarded institutional loyalty more than disciplinary or professional identification.

60 George A. Kelly, "The Autobiography of a Theory," in B. Maher, ed., *Clinical Psychology and Personality* (New York: Wiley, 1969), pp. 46–65, on p. 48.

Kansas. In this context, Kelly revisited the writings about Freudian psychology that he had earlier rejected and found them useful. He experimented in giving patients preposterous explanations about their problems and discovered that it sometimes "worked" – in the sense that it provided some measure of relief to the patient. This did not persuade him that psychoanalysis was correct, but rather that people needed fresh perspectives on their problems and new ways of approaching life's contingencies. After reading the work of J. L. Moreno on psychodrama, Kelly became convinced of the importance of commitment, however temporary, to a given role as people interacted with each other. Thus role playing was a useful therapeutic technique because it allowed people to reformulate their expectations and attitudes toward life.[61]

In November 1939 Kelly traveled to the annual meeting of the Indiana Academy of Science, held at Indiana State Teachers College in Terre Haute, to present a paper entitled "The Person as a Laboratory Subject, as a Statistical Case, and as a Clinical Client." It posed the question of how to reconcile the different conceptions of psychology represented by these three approaches to the individual. This riddle grew directly out of Kelly's variegated interests in physiological psychology, statistics, and psychopathology, as well as his experience in juggling the roles of teacher, researcher, and clinician at Fort Hays Kansas State College. It also reflected general trends within the discipline that had led to the emergence of experimental, correlational, and clinical studies as major lines of inquiry. Kelly did not provide a solution to the puzzle he presented, but predicted that "the definition of the science of psychology eventually will be written by those who are not psychologists."[62]

---

61 Kelly, "Autobiography of a Theory." See also Paul F. Zelhart and Thomas T. Jackson, "George A. Kelly, 1931–1943: Environmental Influences on a Developing Theorist," in Jack Adams-Webber and James C. Mancuso, eds., *Applications of Personal Construct Theory* (Toronto: Academic Press, 1983), pp. 137–154, which contains some biographical details but little analysis.
62 George A. Kelly, "The Person as a Laboratory Subject, as a Statistical Case, and as a Clinical Client," *Proceedings of the Indiana Academy of Science,* 1939, *48:*186. Kelly's self-reported fields of interest appear in *American Men of Science,* 6th ed., 1938.

2

Mobilizing for World War II
From National Defense to Professional Unity

In May 1939 about fifty psychologists attended a reunion in New York City to reminisce about their service in the First World War. Many prominent psychologists were on the guest list, including James Angell, Walter V. Bingham, Truman Kelley, Beardsley Ruml, Walter Dill Scott, Edward Thorndike, Leonard Thurstone, John B. Watson, and Robert Yerkes. They gathered to celebrate the twentieth anniversary of their demobilization as members of the Committee on Classification of Personnel, which had directed the army's personnel system during the war, and to congratulate their former commanding officer, Colonel Walter Dill Scott, who was retiring after two decades as president of Northwestern University.

The event was organized by Walter Bingham, the former executive secretary of the committee who was currently serving as president of the New York State Association for Applied Psychology. Bingham had been a vigorous proselytizer on behalf of applied psychology since the Great War. No doubt aware of the increasingly ominous political situation in Europe, he used the reunion to start rebuilding the military psychology network. One colleague had suggested that a military representative be invited so that "a new tie can be forged which will be the [start] of putting the services of some of the old crowd at the disposal of the country, should the emergency arise." Bingham followed up, and the army sent a lieutenant colonel from the Adjutant General's Office to the gathering.[1]

1 Kendall Weisiger to Bingham, 6 May 1939; WVB/10/May 1–16, 1939; Bingham to E. S. Adams, Adjutant General, 8 May 1939; WVB/10/May 1–16, 1939; William C. Rose to Bingham, 26 May 1939; WVB/10/May 17–31, 1939. Bingham later provided a rather disingenuous account of the occasion: "[We] joined in jovial reminiscence until Colonel William C. Rose, speaking for the Adjutant General, drew attention to the cloud discernible on the European horizon. Within a few short months Hitler marched into Poland. Anticipating what might come to pass, several of us then brushed the dust from contacts with the Army and Navy. . . ." Walter Van Dyke Bingham, in *History of Psychology in Autobiography* (Worcester, Mass.: Clark University Press, 1952), vol. 4, pp. 1–26, quote on p. 22.

Although the reunion of the Committee on the Classification of Personnel did provide an opportunity for psychologists to reflect on the past and lay plans for the future, most members of the "old guard" from World War I were exactly that – too old to take an active part in the coming war. That group had led psychology through two decades of growth and was now being supplanted by a new generation who had come of age during the depression. With a few notable exceptions, the rule that war was a job for younger men would prove to be true again.

In September 1939, four months after the New York reunion of World War I psychology veterans, the American Psychological Association and the American Association for Applied Psychology held their joint annual meeting. The meeting convened on September 4, the day after Britain and France declared war against Germany following Hitler's invasion of Poland. As the news sunk in, both the APA and the AAAP quickly appointed emergency committees to mobilize the profession for national service, and by February 1940 they had joined forces and merged into a single committee.

Before long a national Conference on Psychology and Government Service was being planned. Prior to the meeting, held in Washington, D.C., in August 1940, letters to American psychologists were circulated urging them to remember the profession's accomplishments in the First World War. The "Report of the Psychology Committee of the National Research Council," produced in 1919 by Robert Yerkes, was suggested as an "excellent credential" to present to officials wondering about the possible usefulness of the discipline. Another letter was signed by Bingham, who, as chairman of the newly formed Committee on Classification of Military Personnel, reviewed the situation regarding the use of psychologists in standardizing and administering testing procedures. He noted that psychologists were now in a much better position than they had been in 1917, when they had "to persuade the Army and Navy to give our young profession a chance."

> Today, in the event of grave emergency we are certain to be asked to help on a wide front, bring our technical knowledge, ingenuity and common sense to bear on problems of public attitude, morale, leadership, proper treatment of conscientious objectors, psychological examination of offenders, problems of training and of personnel management in war industries, and on numerous other kinds of tasks, in addition to those arising in connection with the classification and allocation of enlisted and commissioned personnel.[2]

Bingham was extrapolating from the experience of World War I, when personnel work provided the most valuable arena for the exercise of psychological expertise.

---

2 Walter R. Miles to Chairmen of Departments of Psychology and to Members of the American Psychological Association, 1 July 1940; W. V. Bingham to Chairmen of Departments of Psychology and to Members of the American Psychological Association, 1 July 1940; APA/G-8/Yerkes Committee, Miscellaneous Reports.

This war, he predicted, with its potential for even more massive mobilization, offered correspondingly greater opportunities for the profession.

As efforts to coordinate the mobilization of American psychologists on a national scale were initiated, Bingham (1880–1952) and Yerkes (1876–1956) were working behind the scenes to promote the potential military applications of psychology. Both men had played key roles in the First World War. Yerkes had headed the army's intelligence testing program, while Bingham had helped administer its extensive aptitude testing program. The war had been a defining event in their professional lives, having convinced them that applied psychology should play a major role in managing individual behavior and social life. The advent of the Second World War provided them with another chance to demonstrate the utility of psychology through national service, and they were determined to take full advantage of the opportunity.

Bingham, self-employed as a consulting psychologist, stepped up his efforts to get involved in war preparations following the reunion of the Committee on the Classification of Personnel in the spring of 1939. Working through the AAAP, he served as an intermediary between the army and the National Research Council (NRC) to help set up an advisory committee on military personnel problems. Bingham was asked by a military official to draft a letter for the War Department to send to the NRC requesting assistance. Bingham was then among the recipients of the letter he had helped to write. Thus he was responsible for shaping both the approach and the response of the negotiating parties. This sort of bureaucratic maneuvering in order to follow official procedures was typical of the close cooperation that often developed between psychologists and their clients. Soon after this episode Bingham landed a spot on the emergency committee organized by the APA.[3]

By early 1940 Bingham was acting as an unofficial advisor to the War Department on personnel matters. The Adjutant General's Office (AGO) wanted to construct a set of examinations to sort the growing numbers of draftees and enlisted men coming into the army. Drawing on his wide contacts among applied psychologists, Bingham recommended several of his colleagues to aid the AGO in developing its personnel system.[4] He also emerged as a point man for government contacts on the APA emergency committee. His conversations with government and military officials had convinced him that a potentially large market for the services of psychologists existed in connection with emergency preparations. In contrast to the previous world war, when American psychologists had to convince their new clients of their usefulness, Bingham believed that in the event of total mobilization the demand for psychological experts would outstrip the available supply of psychologists.[5]

3 Bingham to Carl Guthe, 26 October 1939; Bingham to Rose, 26 October 1939; WVB/10/October 1939. Willard Olson to Bingham, 24 November 1939; WVB/10/November 1939.
4 Bingham to H. C. Holdridge, AGO, 2 April 1940; WVB/10/April 1940.
5 Bingham to Walter R. Miles, 14 March 1940; WVB/10/March 1940.

Before long Bingham gained the major role he was seeking. Appointed chairman of the AGO's Committee on the Classification of Military Personnel in April 1940, by August he had acquired official status as the army's chief psychologist and the military rank of colonel. As head of the army's personnel psychology program, he became deeply immersed in planning and implementing classification procedures. Bingham's operation became a key venue for the development of applied psychology during the war.

While Bingham was establishing a niche in the federal bureaucracy, Yerkes was trying to gain a hearing for psychology among high-level officials in scientific, governmental, and military circles. Although he was close to retirement age, he maintained a vigorous schedule, and as war approached, he spent more and more time away from his research at the Yale Laboratories of Comparative Psychobiology.

Soon after war began in Europe in 1939 Yerkes corresponded with prominent MIT physicist Karl Compton of the War Resources Board. He suggested that the board, designed as a civilian planning agency, appoint a consultant to deal with the selection and deployment of manpower. At first Compton misunderstood Yerkes's recommendation and replied that military personnel matters were to be handled by the armed forces. Yerkes explained that he was speaking in overall terms, proposing that the human resources of the nation be mobilized as effectively as its material ones. His suggestion, like so many in those days, was duly filed and forgotten.[6]

This minor exchange between Yerkes and Compton sheds light on some of the obstacles facing psychologists as they tried to join federal science mobilization efforts. Top advisors like Compton were drawn almost exclusively from the ranks of physical scientists and engineers located at elite eastern universities. They were often unconcerned and sometimes ignorant about "softer" sciences such as psychology. Furthermore, their tasks were restricted to the tangible materials of warfare – weapons development, communications equipment, transport vehicles, and the like. Psychology, even though it was represented in the National Research Council, could be easily lumped together with the social sciences and their unfavorable associations with government planning and the New Deal.[7]

6 Yerkes mentioned Rutgers University president Robert Clothier as a possible candidate; he was a former businessman and had worked with Yerkes in World War I. Compton to Yerkes, 6 September 1939; Yerkes to Compton, 15 September 1939; Compton to Yerkes, 22 September 1939; Yerkes to Clothier, 22 September 1939; Clothier to Yerkes, 26 September 1939; RMY/94/1789.

7 The tangled relationships among social science, natural science, and the federal government during the 1930s have not been fully explored by scholars. It seems clear that Compton and his colleagues (e.g., James B. Conant, Frank Jewett, Vannevar Bush, etc.) were political conservatives, uncomfortable with if not opposed to the social programs of the New Deal. In contrast, social scientists such as Charles Merriam, Beardsley Ruml, Wesley Mitchell, and others advocated and administered various government social initiatives. The failure of the Science Advisory Board, comprised of physical scientists, within

Yerkes was quite familiar with this bias among members of the scientific establishment and worked assiduously to overcome it. For instance, he continually reminded the leaders of the NRC and of its parent body, the National Academy of Sciences, of the usefulness of military psychology. He told the chairman of the National Research Council about his own willingness to serve, saying, "the concentrated experience which resulted from my period of service in the Army and in the Council I consider precious. It is possible that it and my lifelong work as a student of problems of behavior and human engineering may render me more useful for other tasks than those which happen to engross me in my academic laboratory."[8]

By early 1941 the mobilization of American scientists was already well under way when Yerkes chided Frank Jewett, president of the National Academy of Sciences, for failing to mention the human sciences in a recent article. Yerkes followed up with a copy of his memo on military psychology and offered to help set up a coordinating committee analogous to the Committee on Scientific Research.[9] By then, however, the federal Office of Scientific Research and Development had been established, without explicit provision for psychology.

In addition to scientific leaders, Yerkes lobbied government officials to include psychology in their plans. In early 1941 he orchestrated a major effort to gain the ear of the secretary of war, Henry Stimson. After having persuaded retired Yale president James Angell and others to prepare the way with letters of introduction, Yerkes sent Stimson a copy of his memo on military psychology. Although psychologists had already found a niche in the army's personnel testing program under Bingham, Yerkes recommended that their work be expanded to include military training, the enhancement of morale, and the treatment of deviancy. He also advocated the formation of a training school in military psychology similar to the one that had operated during the First World War in order to ensure an adequate supply of properly prepared psychologists. Yerkes discussed his plans with various military officers, including Brigadier General Frederick H. Osborn. Osborn, a prominent eugenist and public servant, was chairman of the Joint Army and Navy Committee on Welfare and Recreation. Highly supportive of social science, he received an army commission and became head of the Information and Education Division, which conducted extensive research on soldiers' attitudes and morale.[10] But Yerkes was unable to convince military leaders to adopt a comprehensive plan

the social-science-oriented National Resources Planning Board is a case in point. For a discussion of some of the issues, see: Robert Kargon and Elizabeth Hodes, "Karl Compton, Isaiah Bowman, and the Politics of Science in the Great Depression," *Isis,* 1985, 76:301–318. See also A. Hunter Dupree, *Science in the Federal Government* (Baltimore: Johns Hopkins University Press, 1986; orig. pub. 1957), pp. 344–368.

8 Yerkes to Ross Harrison, 11 June 1940; RMY/95/1800.

9 Yerkes to Jewett, 7 January 1941; Yerkes to Jewett, 17 February 1941; RMY/95/1800.

10 See Samuel A. Stouffer et al., *Studies in Social Psychology in World War II,* 4 vols. (Princeton, N.J.: Princeton University Press, 1949–1950).

for the utilization of psychology and it continued to be incorporated into mobilization efforts on an ad hoc basis.[11]

Mobilizing the Profession

The collective response of psychologists to the national emergency was initially channeled through existing organizations. At the September 1939 joint annual meeting of the American Psychological Association and the American Association for Applied Psychology, program chairman Walter Miles made some impromptu remarks at the opening session acknowledging the widespread concern over the international situation. But the meeting adhered to its scheduled program of scientific papers and discussion. An unexpected opportunity for national service arose when psychologist Dean Brimhall, research director for the Civil Aeronautics Authority, announced that Congress had appropriated nearly $6,000,000 for the training of approximately fifteen thousand civilian airplane pilots. A portion of the funds was earmarked for research, and psychologists were needed to conduct studies on pilot selection and training. After the meeting the project was taken up by the Division of Anthropology and Psychology of the National Research Council, and a Committee on Selection and Training of Aircraft Pilots was appointed.[12]

The national associations took additional steps to organize at the meeting. APA President Leonard Carmichael (1898–1973), a prominent animal experimenter and president of Tufts University, appointed an emergency committee to prepare the profession for national defense. Yale experimentalist Walter Miles was chair of the group of thirteen psychologists drawn from leading psychology departments.[13] The AAAP soon followed suit and appointed its own emergency committee. It had already circulated a tentative proposal to expand the use of psychological techniques in the War Department a few months earlier. And shortly after the annual

11 Yerkes to Angell, 9 January 1941; Yerkes to Stimson, 11 January 1941; Yerkes to Stimson, 19 February 1941; RMY/95/1792. Yerkes to Osborn, 17 February 1941; Osborn to Yerkes, 16 March 1941; Yerkes to Osborn, 5 April 1941; RMY/95/1801.

12 The committee's original name included "Civilian Pilots" instead of "Aircraft Pilots." Psychologist members were: J. G. Jenkins (chairman), H. M. Johnson, H. S. Liddell, R. A. McFarland, W. R. Miles, L. J. O'Rourke, C. L. Shartle. Walter R. Miles, "Preparations of Psychology for the War," unpublished paper (c. 1945); APA/G-8/Yerkes Committee, Miscellaneous Reports. The CAA contract to the Committee on Selection and Training of Civilian Pilots was one of the largest administered by the NRC during the war, amounting to a total of $663,500; Rexmond C. Cochrane, *The National Academy of Sciences: The First Hundred Years, 1863–1963* (Washington, D.C.: NAS, 1978), p. 416.

13 Members were: W. R. Miles (chair), M. Bentley, W. V. Bingham, L. Carmichael (ex officio), H. E. Garrett, W. S. Hunter, J. G. Jenkins, H. S. Langfeld, R. A. McFarland, L. J. O'Rourke, D. G. Paterson, A. T. Poffenberger, C. L. Shartle, R. M. Yerkes.

meeting the association sponsored a roundtable in Washington, D.C., on "Possible Psychological Contributions in a National Emergency" that drew together a panel of psychologists along with a number of military and government officials.[14]

Before long the APA and AAAP groups, which had significant membership overlap, recognized their common purpose and joined forces as the Joint Emergency Committee in February 1940.[15] Realizing that closer ties to the federal science establishment were needed, its leaders sought additional legitimacy by reorganizing under the National Research Council.

One of the main vehicles for the pursuit of professional usefulness in the context of national defense was the Division of Anthropology and Psychology of the NRC. The NRC had been formed during the First World War to mobilize the American scientific community. An offspring of the National Academy of Sciences, composed of the country's most eminent natural scientists, the NRC was a quasigovernmental body with a wide-ranging authority derived more from the social status of science than from statutory powers.[16] The Division of Anthropology and Psychology was established after the war, in 1919, with Walter Bingham as its chairman, who in turn had been appointed by James Angell, chairman of the NRC.

Over the years the division had provided a connecting link between psychology and the federal government. At its annual meeting in April 1939 psychologists discussed ways in which they could encourage closer relations between the NRC and the federal government in light of the European situation. Following a failed motion to create a Committee on Tests and Measurement, the representatives authorized a Committee on Public Service to explore the use of psychological tests and other means to aid the government in case of war. The committee, soon renamed the Committee on Selection and Training of Personnel, was headed by John G. Jenkins (1901–1948), chair of the psychology department at the University of Maryland. A dynamic organizer, Jenkins had established an innovative graduate program in applied psychology at Maryland that drew on his wide acquaintanceship among psychologists in business and industry.[17]

14 Richard H. Paynter, "Plans for the Extension of Psychology to the War Department," 17 December 1938, 28 pp.; APA/Yerkes Committee. Lowell S. Selling, chairman, "Possible Psychological Contributions in a National Emergency," 26 November 1939, 20 pp. mimeograph; CML/1/7.

15 There was already significant overlap between the two committees before they merged (e.g., ten members of the APA committee belonged to the AAAP). See W. R. Miles, "Report of the Joint Emergency Committee of the American Psychological Association and the American Association for Applied Psychology," *Psychological Bulletin,* 1940, *37:* 738–741.

16 On the history of the NRC, see Dupree, *Science in the Federal Government;* Daniel J. Kevles, *The Physicists: The History of a Scientific Community in Modern America* (New York: Knopf, 1978); Cochrane, *The National Academy of Sciences.*

17 George K. Bennett, "John Gamewell Jenkins: 1901–1948," *American Journal of*

The Formation of the Emergency Committee in Psychology

As chairman of the joint APA–AAAP emergency committee, Miles suggested that
the NRC sponsor a conference on psychology and government service and invite
representatives from all six national psychology societies. In addition to the APA
and AAAP there were four more specialized groups: the Psychometric Society
(PS), the Society for the Psychological Study of Social Issues (SPSSI), the Society of Experimental Psychologists (SEP), and Section I (Psychology) of the American Association for the Advancement of Science (AAAS). SPSSI and PS were interest groups organized around substantive areas in psychology. The SEP was a
small fraternity of experimentalists. AAAS Section I was a component of the general scientific association in the United States. Sensitive to the issue of proper representation, Miles wrote, "we must manage this in such a way as to amalgamate
the interest and loyalty of the psychologists in all six of our national societies so
that everyone can see and feel that we are acting on a truly national basis for psychology."[18]

Miles helped to lay the groundwork for the conference by circulating two letters addressed to psychology department chairmen and APA members. In one, he
outlined the various committees that had been established to coordinate psychological work and urged psychologists to refer to the record of military psychology
during the First World War. The other letter, signed by Walter Bingham as chairman of the Committee on Classification of Military Personnel, dealt with the development of personnel testing programs.

The Conference on Psychology and Government Service was held in Washington, D.C., in August 1940. The national societies were represented by their presidents and secretaries. The meeting set a pattern of interorganizational cooperation
that persisted throughout the war. With a combined membership of more than three
thousand, these organizations represented the bulk of the country's professional
psychologists. Conferees heard reports from psychologists already involved in
government defense preparations and discussed ways of effectively mobilizing the
entire profession. Discussion led to unanimous agreement to form a central coordinating committee, christened the Emergency Committee in Psychology (ECP),
composed of representatives from each society.[19] With the National Research
Council as its sponsor, the Emergency Committee had a secure base for its work

Psychology, 1948, *61:*433–435; John G. Jenkins, *Psychology in Business and Industry*
(New York: Wiley, 1935); idem, "A Departmental Program in Psychotechnology," *Journal of Consulting Psychology,* 1939, *3:*54–56.

18 Miles to Carl Guthe, 23 July 1940, quoted in Miles, "Preparations of Psychology for the
War," p. 8; APA/G-8/Yerkes Committee, Miscellaneous Reports.

19 Karl M. Dallenbach, "The Emergency Committee in Psychology, National Research
Council," *American Journal of Psychology,* 1946, *59:*496–582.

and access to some of the political resources of the national scientific community. Leading psychologists such as Yerkes, Miles, Carmichael, Hunter, and others were already used to working through the NRC and saw the Emergency Committee as a promising vehicle for the mobilization of psychology.

The Emergency Committee was intended to serve as an advisory body rather than as an operating agency. Acting as a liaison between the psychological profession and the federal government, the committee was to keep psychologists informed about relevant opportunities in military and civilian agencies. Karl Dallenbach (1887–1971), a Cornell experimentalist active in the AAAS psychology section, was selected as chairman. He chose three additional at-large members – Yerkes, Carmichael, and Hunter – while each of the national societies sent one representative. Dallenbach, proud of his First World War rank of captain, interpreted his duties as chairman very conservatively and hewed closely to formal administrative protocols. His cautious leadership style generated controversy later when the war heated up and some psychologists became critical of the slow pace of the committee.[20]

The membership of the Emergency Committee represented most of the major psychology departments in the country. In that way it was a product of the current disciplinary power structure and reflected the existing status quo. While more senior figures such as Hunter and Dallenbach provided leadership, there were many relatively junior psychologists in the group. It provided a place where young professionals from major programs, such as Ernest Hilgard from Stanford or Dael Wolfle from Chicago, could hone their organizational skills and build professional networks. Thus it also provided a means for the emergence of a new generation of leaders in the profession.

The Emergency Committee quickly appointed a number of subcommittees to work on particular problems related to mobilization. Feeling its way at the first meeting, it established subcommittees on perceptual problems, bibliography of military psychology, neurosis, and war experiences and behavior. Without organizational precedents or clear statements of needs from potential clients, the Emergency Committee only gradually developed a distinct operating style. By forming a subcommittee whenever a topic or problem was raised, it served as an inclusive body that could accommodate nearly any special interest, however minor or marginal. Thus the number and variety of subcommittees proliferated quickly, and their heterogeneous nature reflected the diversity of efforts to apply psychology. The subcommittee structure also allowed the ECP to respond quickly to changing wartime conditions. If movement was blocked in one direction, old subcommittees could be abandoned and new ones formed. A few subcommittees proved to be powerful and enduring, such as the Subcommittee on Survey and Planning. But

20 Yerkes, for instance, was highly critical of Dallenbach's leadership. Yerkes to Lewis M. Terman, 9 January 1942; RMY/Terman correspondence file.

real power was shared by a relatively small inner circle composed of members of the psychological establishment and their aides and apprentices.

In the fourteen months prior to Pearl Harbor the Emergency Committee met eight times. Although it was officially constituted as an advisory group, in the absence of any serious competitors for professional leadership it inevitably began to set policy and start programs. Dallenbach's narrow interpretation of its mandate gave way before the pressures for immediate action. He exerted loose control over the group and continued to try to limit its authority. But Yerkes, Boring, Carmichael, and others successfully expanded its functions until it became a virtual "war cabinet" for psychology.[21]

Robert Yerkes and Human Engineering

In April 1941 seven distinguished psychologists read papers at the annual meeting of the American Philosophical Society in Philadelphia. The symposium at the venerable scientific institution was entitled "Recent Advances in Psychology." Each of the participants was asked to address current scientific developments in his research specialty, and the areas of child, Gestalt, abnormal, physiological, educational, and learning were each considered in turn. When the time came for Yerkes, a prominent comparative psychologist, to speak, he deviated from the pattern by offering a presentation on "Psychology and Defense." For him, the prospect of war overshadowed possible interest in the latest research findings. As he outlined the possible contributions the discipline might make to national defense, he urged his colleagues to amplify their mobilization efforts. He was well aware that psychologists were already extensively involved in government war preparations, and that preliminary steps had been taken to organize the profession. Yerkes's presentation carried a distinct message that made it stand out from the usual call to action, however. It contained a detailed blueprint for the deployment of psychology in the war effort that was embedded in a much larger plan for the further professionalization of applied psychology.[22]

By early 1941 Yerkes had distilled his ideas into a concise memorandum outlining a comprehensive plan for military psychology. He classified military resources into two main types, mechanical and human, and claimed that manpower considerations had been neglected in favor of procuring the supplies and equipment necessary for modern warfare. In other words, personnel was as important as matériel in preparing for war. Drawing an explicit analogy to traditional engineering

21 It was dubbed a "war cabinet" by Leonard Carmichael in his review of the book, *Psychology for the Fighting Man,* in *Science,* 1943, *98:*242.
22 Speakers were Arnold Gesell, Wolfgang Kohler, Carney Landis, Karl Lashley, Edward Thorndike, Edward Tolman, and Yerkes. Robert M. Yerkes, "Psychology and Defense," *Proceedings of the American Philosophical Society,* 1941, *84:*527–542.

disciplines, Yerkes argued for increased attention to "human engineering." Human engineering covered all phases of military personnel work, including classification, selection, and training of soldiers; the enhancement and maintenance of morale; and the treatment of malingerers and offenders.[23]

Yerkes attempted to goad military leaders into action by pointing out how the Nazis had based their impressive military personnel system on principles developed by their large corps of military psychologists. Thanks to recent bibliographic reviews as well as firsthand reports by émigrés, American psychologists had ready access to information about German developments.[24] Yerkes complained that the United States military had not built upon the pioneering work that he and others had performed in World War I, and went so far as to add "that what has happened in Germany is the logical sequel to the psychological and personnel services in our own Army during 1917–1918." He recommended the establishment of a separate psychological corps within the armed forces that would place advisors throughout the military bureaucracy, modeled after technical groups such as the Medical Corps or the Chemical Warfare Service. He also proposed that a central research laboratory and training center be created to assure the continuing development of military psychology. In order to foster greater understanding of psychological concepts and techniques among army leaders, Yerkes suggested that courses on military psychology be introduced at the U.S. Military Academy.[25]

Yerkes's plans for military psychology were the opening wedge in a larger campaign to transform the professional role of psychology. He was seeking a greater sphere of action for applied psychology, both within the profession itself and in society. Yerkes's ideas had begun to take shape after World War I, when he and colleagues such as Raymond Dodge formulated a notion of "mental engineering."[26] Extending past the borders of applied psychology, the mental engineering project came to include all aspects of social control. As behaviorism came to the fore, the term "human" was substituted for "mental," and human engineering was portrayed as the analogue of the human sciences. By using the familiar rhetoric of engineering Yerkes was attempting to appropriate some of the social power and cultural

23 Robert M. Yerkes, "Man-power and Military Effectiveness: The Case for Human Engineering," *Journal of Consulting Psychology,* 1941, *5:*205–209.
24 Heinz Ansbacher, assistant editor of *Psychological Abstracts* and a German immigrant, prepared the first systematic review; Heinz L. Ansbacher, "German Military Psychology," *Psychological Bulletin,* 1941, *38:*370–392. He discusses his wartime work in his autobiography; Heinz L. Ansbacher, "Psychology: A Way of Living," in T. S. Krawiec, ed., *The Psychologists,* vol. 2 (New York: Oxford University Press, 1974), pp. 3–49, on 20ff. See also Ulfried Geuter, *The Professionalization of Psychology in Nazi Germany* (Cambridge: Cambridge University Press, 1992).
25 Yerkes, "Man-power and Military Effectiveness," p. 207.
26 Raymond Dodge, "Mental Engineering during the War," *American Review of Reviews,* 1919, *59:*504–508; idem, "Mental Engineering after the War," ibid., 606–610.

authority of the traditional engineering disciplines.[27] To justify the development of a profession of human engineers Yerkes turned to the theory of cultural lag. This notion – that material progress was outpacing social advancement – provided the basic rationale for the analyses of social scientists who provided advice to the federal government in the Hoover administration. In the 1920s cultural lag entered the public vocabulary easily, resonating as it did with widespread anxiety about technology and social change.[28]

Human engineering was necessary, Yerkes argued, because the traditional learned professions could not cope with multiplying social problems. Particularly acute was the gap "between the human needs which are partially met by the physician and those which the clergyman or priest is expected to satisfy."[29] Yerkes did not specify exactly which human needs might fall into this gap. Instead he pointed to the misguided encroachment of both physicians and clergymen into the hypothesized territory of the human engineer, citing examples such as the psychoanalysis of normal individuals and pastoral psychotherapy for the mentally disturbed. Granting few concessions to competing professions, Yerkes staked a wide claim for psychologists-cum-engineers: "Psychology must stand as a basic science for such universally desirable expert services as the guidance and safeguarding of an individual's growth and development, education and occupational choice, social adjustments, achievement and maintenance of balance, poise, and effectiveness, contentment, happiness, and usefulness."[30]

Thus human engineers, or "psychotechnologists," would offer their professional counsel at every stage in the life cycle, from birth through the school years, to job holding, marriage, and beyond. Personal "adjustment" was the catchword as the individual was fitted to his/her social environment, whether the environment was the family, the school, the corporation, the prison, the hospital, or the clinic.[31]

---

27 See JoAnne Brown, *The Definition of a Profession: The Authority of Metaphor in the History of Intelligence Testing, 1890–1930* (Princeton, N.J.: Princeton University Press, 1992).

28 Sociologist William F. Ogburn has been credited with the definition of cultural lag. On social science and public policy before World War II, see Barry D. Karl, "Presidential Planning and Social Science Research: Mr. Hoover's Experts," *Perspectives in American History,* 1969, *3:*347–409; idem, *Charles E. Merriam and the Study of Politics* (Chicago: University of Chicago Press, 1974); Neil J. Smelser, "The Ogburn Vision Fifty Years Later," in N. J. Smelser and D. R. Gerstein, eds., *Behavioral and Social Science: Fifty Years of Discovery* (Washington, D.C.: National Academy Press, 1986), pp. 21–35. Fear of the consequences of cultural lag played an important role in a proposed "research moratorium" in the late 1920s; see Carroll W. Pursell, "'A Savage Struck by Lightning': The Idea of a Research Moratorium, 1927–1937," *Lex et Scientia,* 1974, *10:*146–158.

29 Yerkes, "Psychology and Defense," p. 535.

30 Ibid., p. 536.

31 See Donald S. Napoli, *Architects of Adjustment: The History of the Psychological*

Yerkes was careful not to intrude upon traditional medical turf in his proposals. He emphasized that psychologists should be concerned with the adjustment of normal individuals to life choices rather than with therapy for the abnormal or severely maladjusted.[32] Indeed, he painted a picture of psychology as a professional complement to medicine, dealing with the normal human in a manner analogous to medicine's treatment of the diseased and abnormal. Here Yerkes revealed his characteristic emphasis on psychology as a biological science (hence his preference for the term "psychobiology"), more akin to the natural than to the social sciences. This formulation stressed the "objective" nature of psychological knowledge and masked the social interests motivating his reform program.[33]

The Applied Psychology Panel

In the wake of the attack on Pearl Harbor, psychologists tried again to find a place in the Office of Scientific Research and Development (OSRD), the government's main wartime science agency. Although an indirect link was soon established, it was not until almost two years later, in late 1943, that the Applied Psychology Panel was formed to coordinate psychological activities in the OSRD. By then psychologists had been widely mobilized, and the panel's role was largely confined to monitoring government contracts for military psychology projects conducted by universities and other outside agencies.

The difficulties encountered in the establishment of the Applied Psychology Panel reflected the field's ambiguous status as part of the natural sciences and the domination of physical scientists in the leadership of the American scientific community. The OSRD, along with its contracting arm, the National Defense Research Committee (NDRC), functioned as the main vehicle for the mobilization of the American scientific community during World War II. Utilizing an extensive program of government contracts to academic and industrial researchers, it tapped the scientific and technological resources of the country. Federal contracts for research and development, hardly employed prior to the war, proved enormously useful in the war effort, providing speed and flexibility in meeting changing national needs while at the same time preserving much of the pluralistic structure of American science.

The Office of Scientific Research and Development was directed by Vannevar

Profession in the United States (Port Washington, N.Y.: Kennikat Press, 1981), chap. 2, which rightly notes the pervasiveness of the concept, but overestimates its influence by making it the primary engine of professionalization in psychology.

32 Yerkes, "Psychology and Defense," p. 539. His continuing support for eugenics is clear in this context.

33 See Donna J. Haraway, *Primate Visions: Gender, Race, and Nature in the World of Modern Science* (London: Routledge, 1989), pp. 59–83.

Bush, an engineer with experience and connections in academic, industrial, and government research circles. The agency was charged with developing instruments of warfare. In practice this meant an emphasis on the refinement of existing military hardware and the creation of new weapons. The structure of the NDRC reflected this almost exclusive concern with physical science and engineering. Its twenty divisions and panels were functionally specialized by problem, technique, or subject area.

The domination of physical scientists in the leadership of the American scientific community was reflected in the organization of both the NDRC and the OSRD. Thus it is not surprising that these new agencies made no explicit provisions for psychological research. Despite the fact that psychology was among the disciplines represented in the National Academy of Sciences and the National Research Council, few leaders had been drawn from its ranks. A number of individual psychologists, however, were able to take advantage of federal contracts by linking their work to some of the psychological problems presented by modern weapons systems. Thus researchers in such areas as psychophysics and perception garnered funds from various NDRC divisions.

The origins of the Applied Psychology Panel can be traced to the fall of 1941, when the OSRD requested a survey from the National Research Council on the problem of night vision. Leonard Carmichael, president of the APA the previous year and director of the National Roster of Scientific and Specialized Personnel, chaired the Committee on Human Aspects of Observational Procedures set up to investigate the various problems associated with visual perception under wartime conditions. The committee's main investigator was Charles W. Bray, a researcher in auditory electrophysiology from Princeton, who interpreted his task broadly and gathered additional information on research organizations dealing with psychological factors in classification, training, and equipment. He found some fifty separate groups doing such work, nearly all isolated from each other. Although many were performing essential and useful research, he was concerned that "no effective organization existed to classify the psychologists themselves" and coordinate their work.[34]

Citing the need for better research coordination, Carmichael wrote to Bush in the first months of 1942 suggesting that the OSRD create a psychology section. The two men were already well acquainted. Carmichael was president of Tufts University, where Bush was a member of the board of trustees and his sister served as a dean. Bush was unwilling to do anything concerning psychology unless assurances of high-level military support could be obtained. Following a conference with representatives from the army, navy, OSRD, NDRC, and NRC, a Committee

34 The survey was never published. Charles W. Bray, *Psychology and Military Proficiency: A History of the Applied Psychology Panel of the National Defense Research Committee* (Princeton, N.J.: Princeton University Press, 1948), pp. 8–9, quoted on p. 9.

on Service Personnel was organized in the National Research Council in June 1942 under an NDRC contract. Later that fall the NDRC underwent a major restructuring, expanding its original five divisions into nineteen. But psychology continued to be administered separately through the NRC Committee on Service Personnel.[35]

The civilian membership of the Committee on Service Personnel (CSP) was drawn from the fields of experimental and industrial psychology and psychometrics. Although most of the members were academics, all had had experience in applied research. Carmichael, as chair of the NRC Division of Anthropology and Psychology, was an ex officio member of the committee and acted as the contractor's technical representative for the NDRC. The committee was chaired by John M. Stalnaker, a Princeton faculty member and associate secretary of the College Entrance Examination Board.[36] Charles Bray became executive secretary of the committee and was assisted by John L. Kennedy, a junior faculty colleague of Carmichael's at Tufts. With the exception of Kennedy, the psychologists of the Committee on Service Personnel were around forty years old with at least a decade of professional experience. All were comfortable in performing research for clients outside the university and were already involved in war work.

By the middle of 1943 the work of the Committee on Service Personnel had grown to the point where reorganization was necessary. The Applied Psychology Panel was established in October 1943 to replace the committee. The panel was organized directly under the wing of the Office of Scientific Research and Development, eliminating the layers of bureaucracy that were involved in working under the National Research Council. The Applied Psychology Panel, like the Applied Mathematics Panel, was conceived as an adjunct to the major programs undertaken in the OSRD's regular divisions. The idea behind both panels was that their usefulness cut across the existing organizational categories; many wartime projects had psychological or mathematical dimensions that could be best addressed by this organizational approach. Walter Hunter, an experimental psychologist from Brown University and a member of the Emergency Committee in Psychology, was appointed chief of the panel. Bray and Kennedy remained on the staff as technical aides and were joined by Dael Wolfle from the University of Chicago.[37]

Over the course of the war the Applied Psychology Panel funded twenty major projects that employed a total of two hundred psychologists. Although the research

---

35 Irvin W. Stewart, *Organizing Scientific Research for War: The Administrative History of the Office of Scientific Research and Development* (Boston: Little, Brown, 1948).

36 Other members were: Morris Viteles, University of Pennsylvania industrial psychologist; Clarence Graham, Brown University vision researcher; and George Bennett, director of the Test Division of the Psychological Corporation. Representatives from the armed forces included Walter Bingham, the army's chief psychologist, and P. E. McDowell and F. U. Lake from the navy. Bray, *Psychology and Military Proficiency*, pp. 18–20.

37 Ibid., pp. 23–25.

was usually performed at military field centers around the United States, the contractors were almost always major university psychology departments. Ten universities were involved, along with the Psychological Corporation, the College Entrance Examination Board, and the Yerkes Laboratory of Primate Biology. Under the direction of Hunter's protégé Clarence Graham, Brown University was the largest single contractor. The panel expended a total of $1,500,000, while other units of the National Defense Research Committee spent an additional $2,000,000 on psychological research.[38]

The Office of Psychological Personnel

Although psychology lacked the comprehensive government support provided to other disciplines, such as physics, and did not share in the massive concentration of resources characteristic of "big science" efforts like the atomic bomb project, it did enjoy some advantages that derived from its protean disciplinary identity. If war was fundamentally a matter of conflict among humans, then insofar as psychology was a human science it was conceivably relevant to nearly every aspect of defense mobilization. The human factor was, as always, ubiquitous, and psychologists took pains to remind their potential patrons of that fact.

To a large extent, the psychology community came to depend upon its own devices to manage the demand and supply for trained psychologists during the war. Even though psychologists were included in the comprehensive National Roster of Scientific and Specialized Personnel, that system for tracking and placing technical personnel proved unwieldy and was superseded by other methods organized by discipline or by project.[39] In psychology, the Office of Psychological Personnel (OPP) was established as part of the Emergency Committee's initial efforts to coordinate the mobilization. It became, in effect, a central employment agency for

---

38 The universities that received contracts were: Brown, Harvard, Iowa, Pennsylvania, Pennsylvania State, Princeton, Southern California, Stanford, Tufts, and Wisconsin. For a list of major projects see ibid., p. 38.

39 Psychologist Leonard Carmichael served as director. On its operation see Leonard Carmichael, "The National Roster of Scientific and Specialized Personnel," *Science,* 1940, *92:*135–137; idem, "The National Roster of Scientific and Specialized Personnel: A Progress Report," *Science,* 1941, *93:*217–219; idem, "The National Roster of Scientific and Specialized Personnel: 3d Progress Report," *Science,* 1942, *95:*86–89; idem, "The Number of Scientific Men Engaged in War Work," *Science,* 1943, *98:*144–145. See also Carroll Pursell, "Alternative American Science Policies during World War II," in J. E. O'Neill and R. W. Krauskopf, eds., *World War II: An Account of Its Documents* (Washington, D.C.: Howard University, 1976), pp. 151–162; and idem, "Science Agencies in World War II: The OSRD and Its Challengers," in N. Reingold, ed., *The Sciences in the American Context: New Perspectives* (Washington, D.C.: Smithsonian Institution Press, 1979), pp. 359–378.

the profession. In addition to its main function as a broker between psychologists and government employers, the office performed an integrative role by amassing and distributing information on all professional psychologists, regardless of specialty, professional affiliation, or educational credentials.

The OPP grew out of the Subcommittee on Listing of Personnel appointed in May 1941. Louis B. Hershey, director of the Selective Service, had asked the Emergency Committee for help in finding psychologists qualified to assist local Selective Service Boards in determining the mental capabilities of registrants. Steuart Henderson Britt (1907–1979) of George Washington University, who was trained in psychology and in law, was chosen chairman.[40] By July, Britt had assembled a list of 2,300 qualified psychologists broken down by state and city. Although psychological examiners were used for only a brief period by some local Selective Service Boards, Britt's subcommittee found a receptive audience for their work. The Adjutant General's Office, in charge of army personnel matters, was provided with a list of 1,150 psychologists who were subject to military draft. Not surprisingly, this kind of information proved to be quite useful to a number of agencies, and as more psychologists began writing to Britt concerning government and military jobs "his office soon began to take on the nature of an employment agency."[41]

Becoming overburdened by his part-time duties, Britt drew up plans for an expanded national employment office and presented them at the November 1941 meeting of the Emergency Committee. The idea was taken under advisement and referred to Yerkes for study. After the Pearl Harbor attack the next month and the official declaration of war by the United States, the need for such an office became obvious. In February 1942 Britt was named director of the Office of Psychological Personnel (OPP). It was a semiautonomous agency operating under the aegis of the National Research Council and was funded initially by the American Psychological Association and the American Association for Applied Psychology.

A central personnel office was essential if the entire profession was to be mobilized efficiently. Matching the right man to the right job was an enormous task that was at least partly rationalized by the OPP. As military enlistments grew, psychologists were called up in increasing numbers. The Office of Psychological Personnel tried to ensure that their skills were optimally utilized. Positions of leadership in the federal bureaucracy were easily filled by elite psychologists using the old-boy network, but placement of the rank and file required another method.

Because psychologists were spread throughout the military services and in various government agencies, OPP data were invaluable in understanding the range and extent of their deployment. The OPP sent out questionnaires to as

---

40 Dik Warren Twedt, "Steuart Henderson Britt (1907–1979)," *American Psychologist,* 1980, *35:*850.
41 Dallenbach, "Emergency Committee," p. 515.

many psychologists as they could find, including but not limited to the membership of the major professional organizations. A regular system of follow-up postcards aided the updating of personnel files. By noting the probable induction date of an individual psychologist, Britt could write ahead to the War Department to provide information concerning the special skills and qualifications of the prospective inductee. Thus psychologists increased their chances of being assigned to posts that might make use of their background and training in psychology. In cases where psychologists had already been assigned nonpsychological jobs, the OPP was often able to help secure transfers to more appropriate positions.[42]

The files of the Office of Psychological Personnel were used to fill civilian as well as military positions. Psychologists could be matched with the particular needs of prospective employers, ranging from the routine ("a university instructor") to the exotic ("a man with a special type of experimental training to work on a secret research project"). The OPP was quite busy during its first six months, receiving an average of more than a hundred inquiries a week and sending out nearly as many replies. As director, Britt initiated and cultivated personal contacts with military officers and government officials, and with psychologists stationed in the Washington area. He also edited the "Psychology and the War" section of the *Psychological Bulletin* and gave public lectures, including radio addresses. By May 1942, six months after Pearl Harbor, there were well over a hundred psychologists working in the nation's capital, interacting at various formal and informal meetings.[43]

The work of the OPP underscored the fact that *psychology* could be marshaled for the war effort only insofar as *psychologists* could be mobilized. In other words, it was not abstract knowledge but real individuals who were being called upon to render national service. Given the fluid and dynamic wartime situation, psychologists who might initially have been selected for some position on the basis of their expertise often found themselves in situations where common sense and interpersonal skills were more important. The perceived value of psychology, then, critically depended upon judgments about the ability of the psychologist, not only as a scientist but also as a person.

42 The success of OPP placement efforts was attested by a postwar poll of scientists serving in the military indicating that psychologists were more satisfied with the utilization of their scientific skills than any other disciplinary group. U.S. Army General Staff, *Scientists in Uniform, World War II* (Washington, D.C.: U.S. Army, 1948), pp. 52–53.

43 Steuart Henderson Britt, "The Office of Psychological Personnel – Report for the First Six Months," *Psychological Bulletin,* 1942, *39:*773–793; idem, "Radio Broadcasts on 'Psychologists in the War Effort,'" *Psychological Bulletin,* 1942, *39:*665–669. A May 1942 survey indicated that the number of psychologists working for the government in Washington had doubled in a year, from 61 to 120; Dael Wolfle, "Psychologists in Government Service," *Psychological Bulletin,* 1942, *39:*385–405.

Preparing for the Postwar World

Because of its inclusiveness, the Emergency Committee came to serve not only as a central office for mobilization but also as a forum for debating professional issues. Seeking to place the temporary wartime alliance between psychological science and national service on a permanent footing, Yerkes emerged as a leading advocate for professional reform. Working through the structure of the Emergency Committee, he carved out a semiautonomous base for his activities – the Subcommittee on Survey and Planning – and recruited a group of powerful allies. Their agenda was to yoke the academic and applied wings of the discipline together in a common professional enterprise. They sought to enroll the psychology community in their project by reforming one of its oldest and most powerful institutions, the American Psychological Association.[44]

As we have seen, Robert Yerkes saw World War II as an opportunity to achieve the dream of rational social control through scientific psychology that he and other members of the profession had glimpsed in World War I. After his efforts to convince federal authorities of the need for the systematic mobilization of psychology in the current war had largely failed, he turned his attention toward internal professional reform as a means for accomplishing his objectives.

Within six months after Pearl Harbor, Yerkes was seeking ways to increase the effectiveness of the Emergency Committee. In his eyes, the chairman of the group, Karl Dallenbach, was a major obstacle. Dallenbach, a faithful scientific disciple of his doctoral mentor and predecessor as professor at Cornell, Edward B. Titchener, had chosen to interpret the Emergency Committee's mandate narrowly. He believed that the group should play a strictly advisory role and respond only to official requests for assistance rather than initiate projects of its own. Yerkes, in contrast, was inclined to capitalize on the ambiguity of professional authority caused by the war. He saw the disruption of psychology's normal routines as an opportunity for innovation and chafed under Dallenbach's conservative leadership.

In the spring of 1942 Yerkes was asked by the Emergency Committee to investigate the possibility of holding a conference for long-range planning in psychology. He endorsed the idea and was authorized to select a group "to prepare a report on long-range as well as emergency problems."[45] This became the platform from which Yerkes launched his campaign for professional reform.

Displaying his keen understanding of organizational politics, Yerkes chose the

---

44 The reformation of the APA is treated in greater detail in James H. Capshew and Ernest R. Hilgard, "The Power of Service: World War II and Professional Reform in the American Psychological Association," in *The American Psychological Association: A Historical Perspective,* ed. R. B. Evans, V. S. Sexton, and T. C. Cadwallader (Washington, D.C.: American Psychological Association, 1992), pp. 149–175.

45 Dallenbach, "Emergency Committee," p. 530.

following seven psychologists to serve as conferees: Richard M. Elliott, Edwin G. Boring, Edgar A. Doll, Calvin P. Stone, Alice I. Bryan, and Ernest R. Hilgard. Carl R. Rogers joined the original group slightly later. Yerkes was careful to pick influential individuals from both applied and academic psychology who represented a range of interests in the discipline and who were sympathetic to his cause. His choices were partly constrained by the logistics involved in balancing regional representation against proximity to the East Coast corridor, and by the availability of psychologists who were not already inundated with war work.[46]

The conference group was comprised of representatives of nearly all the major interests – experimental psychology, mental testing, applied psychology, social psychology – in some form. It was heavily weighted toward major academic departments: Yale, Harvard, Stanford, Minnesota, and Columbia. Of the eight members, only three had obtained their Ph.D.s before 1919. But all the males except for Hilgard and Rogers had been involved in the World War I psychology programs. Each member of the group had been highly active in professional organizations, in particular the APA, the AAAP, and to a lesser extent, the SPSSI.

Yerkes had assembled a hardworking and harmonious group of psychologists, dedicated to professional progress yet sensitive to the power of existing interest groups. They gathered at the secluded yet easily accessible campus of the Vineland Training School in southern New Jersey for the first time in June 1942. Meeting for an entire week, the group produced a wide-ranging series of recommendations. The overall thrust of their report emphasized increased planning for psychology, both in wartime and for the postwar period. To that end, they recommended the establishment of a general planning board by the National Research Council independent of the Emergency Committee. Dallenbach, perceiving a threat to the primacy of the Emergency Committee, vetoed the idea and formed the Yerkes group into a Subcommittee on Survey and Planning instead.

The Reformation of the American Psychological Association

The most revolutionary idea proposed by the Yerkes group was that a "central American institute of psychology" should be established "to provide professional services of personnel, placement, public relations, publicity, and publication." The group viewed the wartime Office of Psychological Personnel as a prototype of such an organization and sought to expand its functions and place them on a more permanent footing. They presented a whole set of related recommendations, and suggested that

---

46 Elliott and Boring were longtime friends and professional colleagues who had worked with Yerkes during World War I. Slightly younger, Doll and Stone had also been involved in the World War I program, as predoctoral students. Hilgard had gotten to know Yerkes as a Yale graduate student. Bryan was chosen because of her leadership among women psychologists, whereas Rogers provided a strong link to clinical interests.

they be discussed at an intersociety convention. The Emergency Committee agreed, and arranged for the dissemination of the subcommittee's plans.[47]

Envisioning the postwar role of psychology, the group wrote:

> Psychology as the science of behavior and experience and as a major basis for mental engineering undoubtedly will play an increasingly important role in human affairs. . . . In the new world order its knowledge and skills should be professionalized steadily and wisely so that its applications may keep pace with emerging human needs and demands for personal and social guidance. Foremost among the conditions necessary for the sound and socially profitable maturation of the science and of its technology are: unity of spirit and action; optimal provision for the effective training of psychologists as teachers, practitioners, and investigators; and the creation of such occupational specialties within applied psychology as will satisfy individual and group demands for help in living.[48]

Although the convictions expressed in this paragraph were endorsed by the entire committee, the phrasing was pure Yerkes. His agenda was now attached to a potentially successful instrument, and the process of enrolling the rest of the community had begun.

In September 1942 the Emergency Committee voted "to endorse in principle the development of a permanent service organization for psychology" and called for an intersociety convention. The convention was tentatively scheduled for the spring of 1943, and each of the six societies represented in the Emergency Committee (APA, AAAP, SPSSI, Psychometric Society, SEP, and AAAS Section I) was invited to send five delegates apiece.

Efforts were made to make the convention even more inclusive. The recently formed National Council of Women Psychologists, of which Alice Bryan was a prime mover, successfully petitioned for representation of their 200-member group. The obscure National Institute of Psychology, founded in 1929 and composed of prominent experimentalists, managed to obtain an invitation, despite the fact that the organization existed entirely on paper.[49] In addition to these two

---

47 Dallenbach, "Emergency Committee"; Robert M. Yerkes, Edwin G. Boring, Alice I. Bryan, Edgar A. Doll, Richard M. Elliott, Ernest R. Hilgard, and Calvin P. Stone, "First Report of the Subcommittee on Survey and Planning for Psychology," *Psychological Bulletin,* 1942, *39:*619–630, on p. 629.

48 Yerkes et al., "First Report," pp. 623–624.

49 Incorporated in Washington, D.C., in 1929, the National Institute of Psychology had a membership limit of fifty, composed of active researchers under sixty years of age. Apparently the group's only activity was the election of new members, who were generally well-known experimental psychologists.

groups the department of psychology of the American Teachers Association, an African-American organization, was also invited to send a delegate.[50]

As the plans were being publicized, the modifier "constitutional" mysteriously slipped into the deliberations of the Emergency Committee, and the meeting became known as the "Intersociety Constitutional Convention." Neatly echoing the title of the convention that organized the United States of America, the phrase resonated with psychologists' patriotic sentiments and probably contributed to the sense that a basic restructuring of the profession was appropriate.[51]

The number of delegates from each society was first set at five apiece, until the organizers realized the advantages of proportional representation based on membership size. Some, but not all, of the small groups voluntarily decreased their number of delegates. Thus the Psychometric Society had five representatives, as many as the APA and AAAP, while the remainder had three or fewer. Each society was free to choose its delegates to the Intersociety Constitutional Convention in its own way. The APA conducted an elaborate election for its representatives. From a long list of nominations, a slate of eighteen candidates was drawn up. APA members were instructed to vote for their choices by ranking them. Only 359 ballots were returned from over 700 eligible voters, indicating something less than enthusiastic interest. Five delegates (all former APA presidents) and four alternates were thus elected. In contrast, AAAP president C. M. Louttit was authorized to select representatives from that organization. Survey and Planning Subcommittee

50 Dallenbach, "Emergency Committee," p. 565. The few African Americans who held advanced degrees in psychology were at the margins of the field in the interwar period. Between 1920 and 1939 only sixteen black people earned doctorates in psychology (including educational psychology). Although undergraduate interest in psychology was high in historically black colleges and universities, only four institutions offered an undergraduate major in the subject in 1939. Laboratory facilities and research productivity were likewise lacking, except perhaps at Howard University. In the late 1930s African-American psychologists organized their own professional interest group under the wing of the American Teachers Association, a historically black national education association. Herman G. Canady, "Psychology in Negro Institutions," *West Virginia State College Bulletin, 1939,* Series 26 (3):1–24; Lily Brunschwig, "Opportunities for Negroes in the Field of Psychology," *Journal of Negro Education,* 1941, *10:*664–676; Robert V. Guthrie, *Even the Rat Was White: A Historical View of Psychology* (New York: Harper & Row, 1976), pp. 125–126; F. C. Sumner, "The New Psychology Unit at Howard University," *Psychological Bulletin,* 1935, *32:*859–860; on Canady, see James L. Spencer, *Recollections and Reflections: A History of the West Virginia State College Psychology Department, 1892–1992* (Institute, W. Va.: West Virginia State College, 1994), pp. 21–40.

51 "Preparation for the Intersociety Constitutional Convention," *Psychological Bulletin,* 1943, *40:*127–128. Karl Dallenbach, chair of the Emergency Committee, was perturbed about the implications of the new label, and exaggerated its influence on the subsequent proceedings (Dallenbach, "Emergency Committee," pp. 565–566).

member Ernest Hilgard was an unsuccessful candidate for SPSSI and for APA representative but managed to become a delegate from the obscure National Institute of Psychology. The SPSSI chose three delegates, the National Council of Women Psychologists and the AAAS Psychology Section picked two each, and the department of psychology of the American Teachers Association had one.[52]

Yerkes and his committee members planned carefully for the Intersociety Constitutional Convention. Between March and June, Yerkes sent six letters to the delegates, outlining possible organizational structures and explaining procedures for the meeting. The subcommittee prepared an elaborate forty-six-page *Handbook and Agenda* for the meeting. It contained a statement of the convention's purposes, a discussion of the professional needs of psychologists, and data concerning the Office of Psychological Personnel and existing societies and journals. Its centerpiece

52 Delegates and Alternates to Intersociety Constitutional Convention:

American Psychological Association (5)
Delegates: J. E. Anderson, L. Carmichael, C. P. Stone, R. M. Yerkes, C. L. Hull.
Alternates: S. H. Britt, E. R. Hilgard, H. Woodrow, W. C. Olson.

American Association for Applied Psychology (5)
Delegates: P. S. Achilles, S. H. Britt, A. I. Bryan, E. A. Doll, C. M. Louttit.
Alternates: S. L. Pressey, A. W. Kornhauser, C. R. Rogers, R. A. Brotemarkle, W. C. Trow.

Society for the Psychological Study of Social Issues (3)
Delegates: G. W. Allport, G. Murphy, T. M. Newcomb.
Alternates: E. R. Hilgard, O. Klineberg, G. Watson.

Psychometric Society (5)
Delegates: J. W. Dunlap, H. A. Edgarton, A. P. Horst, I. Lorge, M. W. Richardson.
Alternates: P. J. Rulon, B. D. Wood.

Society of Experimental Psychologists (2)
Delegates: E. G. Boring, R. S. Woodworth.
Alternates: W. S. Hunter, S. W. Fernberger, D. G. Marquis.

National Council of Women Psychologists (2)
Delegates: F. L. Goodenough, G. C. Schwesinger.
Alternates: T. M. Abel, M. A. Bills.

American Association for the Advancement of Science, Section I (2)
Delegates: H. E. Garrett, E. Heidbredder.

National Institute of Psychology (1)
Delegate: E. R. Hilgard.
Alternates: G. R. Wendt, A. T. Poffenberger.

American Teachers Association, Department of Psychology (1)
Delegate: H. G. Canady.

APA/I-6/ICC 1943.

was the presentation of three alternative structures for a national association: a federation of existing societies, the creation of an ideal new society, and the modification of the American Psychological Association. Projected budgets and finances for the various alternative plans were included, along with a suggested timetable for implementing the decisions arising from the convention.[53]

The handbook made explicit reference to the American Constitutional Convention of 1787 in discussing the tension between organizational centralization and decentralization in psychology. In seeking to provide the delegates with useful models of other scientific organizations, the document cited the broad scientific and professional purposes of the American Chemical Society as nearly ideal, requiring only the substitution of "psychology" for "chemistry" and "mental engineering" for "industry."[54]

The handbook also contained some revealing statistics on the organizational affiliations of American psychologists. The figures documented both the expansion of interest groups and the continuing dominance of the American Psychological Association, which was by far the largest and most inclusive society (Table 2.1). Of the three major groups founded in the 1930s – the American Association for Applied Psychology, the Society for the Psychological Study of Social Issues, and the Psychometric Society – only the last drew a significant portion of its members from outside the APA's orbit. Although the AAAP was only one-fifth of the size of the APA in 1942, over one-fourth of the APA's voting members also belonged to the AAAP. The multiple cross-memberships of American psychologists testified to their shared and overlapping interests. But what of those professional psychologists who chose not to belong to these organizations? The APA had long assumed that its membership was practically coextensive with the entire body of the nation's psychologists. Without alternative measures, it was difficult to challenge this assumption. However, the Office of Psychological Personnel, by gathering data on psychologists regardless of organizational affiliation, was able to provide unprecedented statistics on the profession. By early 1943, OPP records revealed a potentially disturbing fact: at least 761 psychologists who were qualified for APA membership were not members.[55]

The Intersociety Constitutional Convention (ICC) was called to order on Saturday morning, 29 May 1943, by Robert Yerkes, chairman pro tempore. Attended by twenty-six delegates from nine national psychological societies, the meeting was

53 Intersociety Constitutional Convention, *Handbook and Agenda,* 1943; APA/K-8/AAAP; Robert M. Yerkes, Edwin G. Boring, Alice I. Bryan, Edgar A. Doll, Richard M. Elliott, Ernest R. Hilgard, and Calvin P. Stone, "Psychology as a Science and Profession," *Psychological Bulletin,* 1942, *39:*761–772.
54 Intersociety Constitutional Convention, *Handbook and Agenda,* pp. 14, 18.
55 S. H. Britt to W. C. Olson, 13 May 1943; APA/K-8/AAAP.

Table 2.1. *Membership in American psychology societies, 1937–1944*

Year	American Psychological Association	American Association for Applied Psychology	Society for the Psychological Study of Social Issues	Psycho-metric Society	Society of Experimental Psychologists	National Institute of Psychology	American Teachers Association, Department of Psychology	National Council of of Women Psychologists
1937	2138			158	50			
1938	2318		374	221	50		10	
1939	2527	411	349	240	50		15	
1940	2739	564	327	236	50		25	
1941	2937	608	291	245	50	43	26	
1942	3231	638	294	232	50	41	29	253
1943	3476	659	242		50		32	261
1944	3806		273					258

Sources: Intersociety Constitutional Convention of Psychologists, *Handbook and Agenda*, 1943, Table 1 (APA/K-8/AAAP/ICC); L. G. Portenier, ed., *The International Council of Psychologists, Inc.* (Greeley, Colo.: ICP, 1967).

held in the Hotel Pennsylvania in New York City. In his opening speech Yerkes compared psychology with the physical sciences and engineering:

> Recent decades have witnessed the rapid transformation of our physical environment by discovery, invention, and the development of engineering skills. The time is ripe for equally innovational changes in human nature, its controls, and expressions. Physical conditions are such that this revolution can happen now. Furthermore, there are signs that it may happen, whatever the attitude of our profession. It is fitting for us to resolve that it shall happen, facilitated to the utmost by our directive energies and our specialized knowledge, wisdom, and skills.

He expressed hope that existing forms of human engineering would be augmented by further attempts "to assure and to increase life's values through services of guidance, direction, counseling and enlightenment."[56]

Like Moses leading the chosen people to the promised land, Yerkes pictured a utopian future for the profession:

> The world crisis, with its clash of cultures and ideologies, has created for us psychologists unique opportunity for promotive endeavor. What may be achieved through wisely-planned and well-directed professional activity will be limited only by our knowledge, faith, disinterestedness, and prophetic foresight. It is for us, primarily, to prepare the way for scientific advances and the development of welfare services which from birth to death shall guide and minister to the development and social usefulness of the individual. For beyond even our wildest dreams, knowledge of human nature may now be made to serve human needs and to multiply and increase the satisfactions of living.

Yerkes concluded his evangelistic address with a benediction: "With you, my fellow psychologists, I pray that from this, the first Intersociety Constitutional Convention of our profession, wisdom may flow like a mighty river of enlightenment and good will."[57]

Permanent officers were elected, including Edwin Boring as chairman.[58] In his opening statement, Boring characterized the problem facing the convention in terms of the conflict between federal authority and states' rights: "Our dilemma concerns the question as to how far certain privileges of interest-groups should be delegated to central authority, and how far they should be retained by these

---

56 Intersociety Constitutional Convention, condensed transcript, 29–31 May 1943; APA/I-6/ICC 1943, pp. 1–2.

57 Ibid., p. 2.

58 They were Edwin Boring (chairman), Ernest Hilgard (vice-chairman), Alice Bryan (secretary), and Edna Heidbredder (vice-secretary). All except Heidbredder had been members of the Survey and Planning Subcommittee.

groups. . . ."[59] By extending the rhetoric of the U.S. constitutional convention, Boring helped to reinforce the perception of the convention as a new beginning for psychology, an opportunity to recast the foundations of professional activity.

The delegates agreed to follow the tentative agenda set before the meeting. They spent the morning discussing the topic "Should the scientific and technological aspects of psychology be developed together or separately?" and describing their aspirations for the convention. After lunch, Boring summarized the morning's discussion under four major points. All agreed that pure and applied work (science and technology) were "inextricably related," but that a "professional attitude" separated applied from academic psychologists. As a consequence, professional psychologists were more concerned with public relations and ethical issues of practice. It was felt that training should encompass both scientific and professional aspects of psychology. Finally, there were questions concerning the definition of the scope of the field. As Boring put it, "broadening the scope of psychology brings with it both advantages and disadvantages."[60]

After this general discussion of the aims of the convention, committees were appointed to consider three possible courses of action: to form a federation of existing societies in order to accomplish common objectives; to create an ideal new association without regard to existing groups; or to modify the bylaws of the APA so it could function more effectively. The committee reports were heard the next morning, and the plans for revising the constitution of the APA generated by far the most discussion.[61] Federation appeared to be an ambiguous and vague solution, and the ideas for an ideal society could be easily incorporated into plans to reorganize the APA. By remaking the structure of the APA, psychologists could have their cake and eat it too, attaining the prestige of an important and venerable society and enjoying the efficiency of a revamped organization. There was little reason or incentive to destroy the APA. American psychologists continued to look to it for leadership and relied on it for scientific legitimation. Its fiftieth anniversary celebration in 1942 had brought in its wake a series of articles and reminiscences that lauded psychology's rich heritage and emphasized its continuing relevance to the present.[62] The dislocations of war underscored the sense that an era was departing, that members of an earlier, pioneering generation were passing the torch to their descendants who were working in a different set of historical circumstances. The presence of septuagenarian Robert S. Woodworth at the convention as a delegate from the Society of Experimental Psychologists can be understood in this

59 Intersociety Constitutional Convention, condensed transcript, p. 3.
60 Ibid., p. 4.
61 For details see Capshew and Hilgard, "The Power of Service," pp. 163–164.
62 See Edwin G. Boring, "The Celebrations of the American Psychological Association," *Psychological Review,* 1943, *50:*1–4.

context. As the embodiment of the catholic "Columbia style," he was a vital link to the early days of American psychology. Among the few surviving members of his generation, he was nearly alone in remaining professionally active.

Nearly all of the second day of the meeting was devoted to working out how to reorganize the APA. Each society representative was given the opportunity to react to the proposal and to bring up any grievances against the APA. Widespread dissatisfaction was expressed over the lack of voting privileges for APA associate members and the consequently restricted chances for participation of younger psychologists. Gladys Schwesinger, speaking for the National Council of Women Psychologists, felt that female psychologists also had inadequate opportunities for participation in APA affairs. In a similar vein, Herman Canady, representative of the black American Teachers' Association, criticized the neglect of black institutions in psychology by the APA.[63] Former APA president Clark Hull added a note of caution to the proceedings, saying, "I want to register my belief that railroading does sometimes occur in the APA especially when controversial issues are presented as unanimously approved without discussion."[64]

In addition to the general problems of representation in the APA, special-interest-group delegates registered specific complaints. Psychometric Society members Phillip Rulon and Marion Richardson complained that their joint annual meeting programs with the APA had suffered because of poor planning by APA organizers.[65] SPSSI secretary Theodore Newcomb, supported by Gordon Allport, felt the APA had not been active enough in protecting the political freedom of psychologists, citing the recent congressional investigation of Columbia psychologist Goodwin Watson.[66] Edgar Doll stated the basic problem succinctly: "The essential failure of the APA to recognize emergent groups is the primary objection to the APA." He contended that if the APA had been more responsive to the social and professional interests of psychologists, separate societies would never have been formed. Psychologists wanted to belong to a unified discipline, and the overlapping memberships between the APA and the other societies evidenced this desire.[67]

63 Canady articulated the interests of black psychologists in a manner similar to the position of the National Council of Women Psychologists, saying: "We are interested primarily in psychology. We don't think of ourselves as negro psychologists. We are simply psychologists who by accident happen to be negroes." Intersociety Constitutional Convention, condensed transcript, p. 17.

64 Ibid., pp. 12–14.

65 This complaint was made despite the fact that the Psychometric Society did not have enough papers to arrange a program for their 1941 meeting, and barely enough papers for the previous meeting. Jack W. Dunlap, "The Psychometric Society: Roots and Powers," *Psychometrika,* 1942, 7:1–8.

66 Benjamin Harris and S. Stanfeld Sargent, "Academic Freedom, Civil Liberties, and SPSSI," *Journal of Social Issues,* 1986, *42* (1):43–67.

67 Intersociety Constitutional Convention, condensed transcript, p. 16.

Steuart Britt of the AAAP indicated that if the APA were to be restructured to include the functions of the AAAP and other groups, the AAAP would go along. Anderson, clearly in favor of restructuring the APA, candidly agreed with the complaints against it. Reminding the group of their unprecedented authority, he said: "Any recommendations which come from such a group as this Convention will be virtually mandatory since we represent different degrees, types, and shades of opinion within the organization."[68] Boring summarized the discussion:

> The Chair thinks it was said that the APA, while seeming to be continued in the proposed plan, would be so altered that it would be incorporated with its sections becoming Divisions of the new APA; that the SPSSI presumably would be made a Division; that Section I [Psychology, AAAS] would be unaffected; that the other societies which have been mentioned [Society of Experimental Psychologists; National Institute of Psychology] would seem to be left outside insofar as they represent interest groups. Additional interest groups would grow up, and existing groups might go out of existence.[69]

As the afternoon session drew to a close, Yerkes tried to redirect the discussion back to the concrete details of revising the APA bylaws. Allport successfully proposed to extend further the APA's stated purpose of advancing psychology as a science and a profession by adding the phrase "and as a means of promoting human welfare."[70]

On the third and final day of the convention conferees finished outlining the basic structure of the new society as follows. The certificate of incorporation was to be reworded to state: "The object of this society shall be to advance psychology as a science, as a profession, and as a means of promoting human welfare." Two classes of membership were proposed: fellow (equivalent to APA member) and member (equivalent to APA associate). In addition, divisions could have similar separate categories for nonmembers if they desired. Divisions, organized around interest groups, required a minimum of fifty members and would be the basis for representation on the council. The initial divisions would be: General Psychology, Clinical, Educational, Business and Industrial, Consulting, Psychometric Society, and SPSSI. New divisions could be started with a petition by at least fifty members and old ones disbanded because of low membership or lack of interest. Geographical branches, such as the regional psychological associations, and other societies would be considered for affiliation with the APA. The Council of Representatives was to be the major governing body, with proportionate representation from each division and with members-at-large elected by the entire association. Provision was made for a representative of black universities and colleges. Two new offices were added: a president-elect and a full-time executive secretary. The

68 Ibid., p. 15.   69 Ibid., p. 16.   70 Ibid., p. 17.

board of directors would consist of the executive officers (president, president-elect, recording secretary, and treasurer) and five members of the council. The creation of a Policy and Planning Board to provide overall direction was suggested. A central office would continue to handle APA publications, as well as those of other groups if they wished.[71]

Carmichael, Doll, and Allport drafted a motion to accept the report:

> [We] move that having given careful consideration to various proposals placed before us, this Convention record its decision that the objectives in view can most effectively and economically be achieved through a closer and more organic tie between the reconstituted present national psychological societies and their present affiliates, and that, in view of legal, material, and professional considerations, the name of this national organization should be the American Psychological Association.[72]

The motion was passed.

In the ensuing discussion, the delegates presented various minor amendments to the plan.[73] Boring appointed a Continuation Committee, consisting of Hilgard (chair), Anderson, Bryan, and Doll (Allport and Boring were later elected as members also), to prepare a detailed reorganization plan that incorporated the proceedings and discussions of the ICC. The plan would then be submitted for consideration to the constituent societies through their delegates. Near the close of the convention, Irving Lorge, a Psychometric Society representative, moved that the plan include a statement "that the reorganization shall go into effect when approved by both the APA and AAAP."[74] This was an important procedural move that created a mechanism for ratifying the changes made by the large associations that represented a majority of American psychologists.

Yerkes and his allies orchestrated the ratification of the reforms as adroitly as they had handled their passage through the Intersociety Constitutional Convention. Although Hilgard was the chairman of the Continuation Committee, he leaned heavily on its senior members, especially Boring, who fussed endlessly over the details. Shortly after the convention Boring asked Bryan to circulate the minutes to the delegates, and to prepare notices for the *Psychological Bulletin* and *Science*. He suggested that the notice emphasize the unification of psychology and gloss over the retention of the name of the American Psychological Association as "merely a convenient means." Bryan was more sensitive to the concerns of AAAP

71 Ibid., p. 25.   72 Ibid., p. 20.

73 Among the most important was Britt's suggestion to allow present AAAP fellows to become fellows of the new APA automatically. Yerkes offered a successful motion that allowed the executive secretary an indefinite term of office.

74 Ibid., p. 27.

members and decided not to mention the new organization's proposed name in the public announcements.[75]

By the middle of July a tentative draft of the plan was sent to the delegates and officers of the national psychology societies. The bulk of the twenty-five-page document consisted of bylaws for the "reconstituted" APA. In a brief summary of the convention, the report emphasized the basic principles of organization: the functional autonomy of interest groups and administrative centralization. Adroitly addressing the issue of sovereignty, it claimed that organizational reform could be accomplished

> only by a reconstitution of the present structure of the American Psychological Association in such a way that it include within its expanded structure the functional interests and professional atmosphere of the American Association for Applied Psychology. The Convention recommends, therefore, a *de facto* amalgamation of these two societies under new forms and a *de jure* continuation of the old society, with its appropriate name, its prestige and [its assets].[76]

At their September 1943 meetings, the APA and AAAP governing councils accepted the Continuation Committee's recommendations and appointed a joint committee to manage their implementation. Their action raised a technical point of order: did APA and AAAP leaders have the authority to act without the formal approval of their memberships? This was a procedural question similar to the one raised about the Intersociety Constitutional Convention and its authority to act on behalf of all American psychologists. As before, the issue was glossed over as the reformers continued to rely on the tacit approval of the psychology community.

Hilgard, Bryan, and Anderson continued as members of the new committee, and Hilgard retained his chairmanship.[77] The joint committee called for an advisory mail vote by AAAP and APA members, then a final revision of the bylaws followed by a formal ratification vote. This schedule would permit formal adoption of the reorganization plan by September 1944 and allow time to elect new officers and to make the organization fully operational by the following year. These recommendations were widely publicized, along with commentary.[78]

---

75 E. G. Boring to A. I. Bryan, 2 June 1943; Bryan to Boring, 9 June 1943; EGB/Bryan 1942–43.

76 Continuation Committee, "Report and Recommendations of the Intersociety Constitutional Convention of Psychologists to the Societies Represented" [July 1943], pp. i–ii; APA/I-5/ICC 1943.

77 Ernest R. Hilgard to E. G. Boring, 5 September 1943; EGB/He-Hi 1942–43.

78 "Recommendations of the Intersociety Constitutional Convention of Psychologists: I. Statement of the Joint Constitutional Committee of the APA and AAAP," *Psychological Bulletin,* 1943, *40:*621–622; "II. Statement by the Continuation Committee of the

The only significant opposition to the plan was voiced by Harriet O'Shea, who chaired the AAAP Board of Affiliates, which represented ten state groups of applied psychologists. These groups consisted primarily of practitioners involved in the delivery of services; their view of applied psychology contrasted with that of the AAAP leadership, which was inclined to include the academic and research aspects of the field as well. In March 1944, in a letter that was subsequently published, O'Shea circulated a list of reasons against the reorganization, arguing that applied psychology might lose its identity in the process. Bryan and other proponents of the reorganization responded to O'Shea's criticisms rapidly and effectively, both publicly and privately, and any potential damage was averted.[79]

Two mail votes regarding the proposal were taken in 1944; both indicated strong approval. One ballot canvassed opinions of APA and AAAP members, while the other was sent to all American psychologists on the rolls of the Office of Psychological Personnel.[80] The APA and the AAAP officially approved the proposal in September 1944 and set September 1945 as the date for the new organization to begin. Hilgard was named chairman of transition committees to work out constitutional and organizational details.[81]

In the process of organizing for national defense, psychologists found themselves planning for their postwar future. The new American Psychological Association represented the apogee of professional unity in the context of war. Patriotic sentiments found ready expression in efforts to extend the reach of the psychological profession. Although interests of African-American psychologists and women were at least nominally represented by the new APA, by and large the wartime environment did not expand their professional opportunities.

Convention," ibid., 623–625; "III. By-laws Appropriate to a Reconstituted American Psychological Association," ibid., 626–645; "IV: Sample Blank for Survey of Opinion on the Proposed By-laws," ibid., 646–647; W. L. Valentine and J. E. Anderson, "Chart of the Proposed APA Reorganization," ibid., 1944, *41*:41; E. E. Anderson, "A Note on the Proposed By-laws for a Reconstituted APA," ibid., 230–234; J. E. Anderson, "A Note on the Meeting of the Joint Constitutional Committee of the APA and AAAP," ibid., 235–236.

79 Capshew and Hilgard, "The Power of Service," pp. 169–170.
80 Willard C. Olson, "Proceedings of the Fifty-second Annual Meeting of the American Psychological Association, Inc., Cleveland, Ohio, September 11 and 12, 1944," *Psychological Bulletin,* 1944, *41*:725–793; Ernest R. Hilgard, "Psychologists' Preferences for Divisions under the Proposed APA By-laws," *Psychological Bulletin,* 1945, *42*: 20–26.
81 Ernest R. Hilgard, "Temporary Chairmen and Secretaries for Proposed APA Divisions," *Psychological Bulletin,* 1945, *42*:294–296.

3

Home Fires
Female Psychologists and the Politics of Gender

As the American psychology community mobilized for national defense after the Nazi invasion of Poland in 1939, the possible contributions of female psychologists were overlooked or ignored. The exclusion of women became glaringly obvious at the joint annual meeting of the American Psychological Association (APA) and the American Association for Applied Psychology (AAAP) in September 1940, where reports on preparations for war dominated the discussions. Gladys Schwesinger, chair of the AAAP's Section of Consulting Psychology, disgustedly noted that, "as the list of activities and persons rolled on, not a woman's name was mentioned, nor was any project reported in which women were to be given a part." Even worse, there was no promise that things would change. Some of the women protested, but with little effect. Summarily ignoring their professional status, the male leaders informed them that tradition favored the services of men in wartime. Women were supposed to "keep the home fires burning." At best, they could expect to "wait, weep, and comfort one another."[1]

Unwilling to take no for an answer, about thirty of the women met to lobby for a role in mobilization efforts. They promptly confronted Robert Brotemarkle, the AAAP representative to the Emergency Committee, with two pressing questions. Would women be omitted from the National Register of Scientific and Specialized Personnel that was being collated? And would they gain representation on the

---

1 This chapter is based on James H. Capshew and Alejandra C. Laszlo, "'We would not take no for an answer': Women Psychologists and Gender Politics during World War II," *Journal of Social Issues,* 1986, *42*(1):157–180. Gladys C. Schwesinger, "Wartime Organizational Activities of Women Psychologists: II. The National Council of Women Psychologists," *Journal of Consulting Psychology,* 1943, *7*:298–299, quoted on p. 298. One early proposal did address the possible role of female psychologists; see Richard H. Paynter, "Plans for Extension of Psychology to the War Department," unpublished manuscript (c. 1939); APA/G8/Yerkes Committee-Misc. Reports.

all-male Emergency Committee itself?[2] Unfortunately, Brotemarkle could only counsel patience and, in particular, urged the women not to take independent action. Convinced that Brotemarkle would not adequately represent their interests, the group sent Millicent Pond, an industrial psychologist from Connecticut, to plead their case directly before the Emergency Committee. Pond received a sympathetic hearing, but again no action was taken. Instead the committee casually admonished the women "to be good girls . . . and wait until plans could be shaped up to include [them]."[3]

Although the group temporarily acquiesced, as the emergency situation intensified so did their desire for action. Female psychologists realized that, whereas they were told to find volunteer work in local communities as an outlet for their patriotic zeal, large numbers of male psychologists were finding paid employment in the military and federal bureaucracy. At the next AAAP conference, in September 1941, protests against this discrimination escalated. With a rapidly expanding market for applied psychology in the government, the women were determined to assert their established expertise. Brotemarkle met with the group once again. Anxious to stem the tide of discontent, he adamantly denied that any discrimination existed in civilian emergency appointments and pleaded with the women to be patient.[4]

The women reluctantly agreed to wait, but, by now more skeptical of how the situation was being handled, they also organized an informal Committee on the Interests of Women Psychologists to act as an information clearinghouse and to explore possibilities for wartime service. Chaired by Harriet O'Shea, a school psychologist at Purdue University and president of the Indiana Association of Clinical Psychologists, the committee began a letter-writing campaign. Not only were Brigadier General Frederick Osborn, chief of the Morale Branch, and Brigadier General Wade H. Haisilip, assistant chief of staff, approached about military opportunities, but Eleanor Roosevelt, a prominent advocate of women's rights, was informed of their quandary.[5]

When by November the requested report was not forthcoming, the women finally lost patience with the tactics of the Emergency Committee. Throughout the fall of 1941 a group of about fifty women psychologists in New York City had been meeting sporadically to discuss their role in the national emergency. With its high concentration of professional psychologists, New York was perhaps the only place where such an ad hoc group could form. Deciding not to wait any longer for action on the part of the Emergency Committee, on November 11, 1941, thirteen of

2 R. A. Brotemarkle to W. V. Bingham, 9 September 1940; WVB/19/Correspondence 1940.
3 Schwesinger, "Wartime Organizational Activities," p. 299.
4 Schwesinger, "Wartime Organizational Activities"; H. E. O'Shea to R. A. Brotemarkle, 17 September 1941; AIB/NCWP.
5 H. E. O'Shea to Committee on Interests of Women Psychologists, 24 November 1941; AIB/NCWP.

these women met in Alice Bryan's Manhattan apartment to map out strategy for a New York chapter of what they hoped would evolve into a separate and national organization for female psychologists.[6] Using O'Shea's committee, they polled women in both the APA and the AAAP to determine interest and support for such an association. The purpose, they claimed, was less to protest their continued exclusion from mobilization committees and government positions than to promote specific projects for female psychologists. This action marked the first serious break from male dominance of professional activities.[7]

Two weeks later, on November 29, and fully a year after the original request, the Emergency Committee appointed a subcommittee "to investigate the possible activities or services of women psychologists in the national emergency."[8] Whether or not it was prompted by the threat of a splinter group beyond their direct control, the Emergency Committee action came too late to appease female psychologists now bent on a national organization. News of Pearl Harbor a week later only strengthened the women's resolve. On December 8, a call letter went out to local psychologists; a week later approximately fifty women met and voted to officially organize the National Council of Women Psychologists (NCWP).[9]

The National Council of Women Psychologists

The formation of the new group represented the culmination of demographic and professional trends that had been gathering momentum since the 1920s as psychology became divided along gender lines. Although women comprised nearly one-third of the entire profession, they had been almost entirely shut out of leadership roles within the American Psychological Association. The situation was somewhat better in the American Association for Applied Psychology, where women held a few administrative posts but still not in proportion to their numbers.[10] The prevalent sex discrimination encountered by women psychologists was also reflected in the dual labor market that had evolved in psychology. Male Ph.D.s tended to hold higher-status jobs in university and college departments, concentrating on

6 H. E. O'Shea to A. I. Bryan, 14 March 1942; AIB/NCWP.

7 G. C. Schwesinger to H. E. O'Shea, 12 November 1941; AIB/NCWP.

8 Ruth S. Tolman, "The Subcommittee on the Services of Women Psychologists," *Psychological Bulletin,* 1943, 40:53–56, quoted on p. 53.

9 C. P. Armstrong, "The National Council of Women Psychologists," in P. L. Harriman, ed., *Encyclopedia of Psychology* (New York: Philosophical Library, 1946); Schwesinger, "Wartime Organizational Activities."

10 L. J. Finison and L. Furumoto, "Status of Women in American Psychology, 1890–1940, Or on How to Win the Battles Yet Lose the War," unpublished paper, 1980; Mildred B. Mitchell, "Status of Women in the American Psychological Association," *American Psychologist,* 1951, 6:193–201. See also Margaret W. Rossiter, *Women Scientists in America: Struggles and Strategies to 1940* (Baltimore: Johns Hopkins University Press, 1982).

teaching and research. Female Ph.D.s, on the other hand, were usually tracked into service-oriented positions in hospitals, clinics, courts, and schools. Discouraged and frequently prevented from pursuing academic careers, women filled the ranks of applied psychology's low-paid, low-status workers. The few women who managed to gain academic employment were mostly relegated to women's colleges, and to university clinics and child welfare institutes linked to departments of psychology and education.[11]

The limited opportunities typically encountered by women psychologists are strikingly illustrated in the career paths of Alice Bryan and Georgene Seward, two Columbia University graduate students in the 1920s. Alice Bryan (1902–1992) first entered Columbia through its extension division in 1918 to pursue a two-year secretarial course. In 1921 she became an instructor in the Extension Division of the United YMCA Schools and taught correspondence courses in advertising to war veterans for the next eight years. After her marriage in 1924 she enrolled as a part-time student at Columbia, receiving a B.S. in psychology in 1929 and a Ph.D. in 1934. Despite her academic credentials, for the next five years Bryan held a series of part-time jobs, mixing teaching, research, testing, and consulting until she finally obtained full-time employment as a psychologist in Columbia's School of Library Service in 1939. Although she retained her professional identification with psychology throughout World War II, her career was permanently diverted into library science.[12]

In contrast to Bryan, Georgene Seward (1902–1992) followed the academic fast track in her graduate career at Columbia. She received a B.A. in 1923 after three years of study and a Ph.D. in psychology in 1928. Her colleagues at Columbia included faculty members Leta and Harry Hollingworth, Gardner Murphy and his student wife, Lois Barclay, and fellow students Margaret Mead and John Seward, whom she married in 1927. The professional disparity often encountered in dual-career marriages helped to sensitize her to gender issues. While her husband taught at Columbia, Seward's career floundered on the lower rungs of the academic ladder for a long time. Following seven years at the rank of instructor at Barnard College, Seward moved to Connecticut College for Women in 1937, where she remained an assistant professor through the war years. Frustrated at being consigned

11 James Capshew, Richard Gillespie, and Jack Pressman, "The American Psychological Profession in 1929: Preliminary Data and Results of a Prosopographical Study," unpublished paper, 1983; Samuel W. Fernberger, "Academic Psychology as a Career for Women," *Psychological Bulletin,* 1939, *36:*390–394; Gladys D. Frith, "Psychology as a Profession," *Women's Work and Education,* 1939, *10*(3):1–3; Finison and Furumoto, "Status of Women."

12 Alice I. Bryan, "Autobiography," in Agnes N. O'Connell and Nancy F. Russo, eds., *Models of Achievement: Reflections of Eminent Women in Psychology* (New York: Columbia University Press, 1983), pp. 68–86.

to what she considered an institutional purgatory, Seward nonetheless struggled doggedly toward the academic mainstream in psychology.[13]

Although graduate education was imbued with an ideology that stressed psychology's identity as an academic discipline based on experimental research, the growing demand for workers in service roles enabled the development of centers of applied training such as Columbia Teachers College and the University of Minnesota. Partly because the lower status of applied work was widely accepted, women gained unusual access to graduate training in this area. Columbia, with its large program and supportive faculty, was perhaps the best place for a woman to pursue graduate education in psychology in the interwar period. In a 1943 survey of active professional psychologists, Columbia had granted 144 Ph.D.s in psychology to women, fully 22% of the female doctorates. At institutions with a strong experimental orientation, however, women fared poorly. Harvard University, where only five women received the Ph.D. (less than 1% of the total), was perhaps the most inhospitable, as the story of Mildred Mitchell suggests.[14]

Mildred Mitchell (1903–1983) was a high school teacher when she entered Harvard summer school in 1926 at the urging of a friend; soon afterward she decided to stay and study for a doctorate in psychology. Although Mitchell found some encouragement for her interests in clinical and abnormal psychology through courses with Morton Prince of the Harvard Psychological Clinic and Frederic L. Wells at the Boston Psychopathic Hospital, she also encountered deeply entrenched sex discrimination. Since women were not allowed to enroll officially as Harvard students, Mitchell was forced to register at Radcliffe and pay her tuition there, despite the fact that all her classes were taken with Harvard faculty members. In her autobiography Mitchell pointedly described her treatment by Edwin Boring, who seemed to delight in erecting petty barriers for female students. For instance, Boring not only refused Mitchell a key to the psychology laboratories, so that she had to wait for "some man with a key" in order to work in the evenings; he also insisted that, when she did work at night with a male advisor, a chaperone be present. The department library and Widener Library were likewise closed to women in the evenings. Boring further impeded Mitchell's studies by barring her from the psychology colloquia because of her sex. He finally provided the catalyst for her decision to leave after she received her M.A. by disparaging Mary Whiton Calkins's refusal to accept a Radcliffe Ph.D. for doctoral work completed under William James at Harvard.[15]

13 Gwendolyn Stevens and Sheldon Gardner, eds., *The Women of Psychology,* vol. 2 (Cambridge, Mass.: Schenkman, 1982), pp. 139–143.

14 Alice I. Bryan and Edwin G. Boring, "Women in American Psychology: Statistics from the OPP Questionnaire," *American Psychologist,* 1946, *1*:71–79.

15 Mildred B. Mitchell, "Autobiography," in O'Connell and Russo, *Models of Achievement,* pp. 120–139.

Mitchell continued her graduate studies at Yale University, where her experience was more positive: she was immediately given a private office, along with keys to the library and laboratories, and was a welcome participant in the department colloquia. Mitchell found a helpful advisor for her hypnosis research in Clark Hull, whose arrival at Yale coincided with her own, and she completed her dissertation in 1931. Nevertheless, for the next decade Mitchell held a succession of low-paying, low-status jobs in mental hospitals and clinics.[16]

Each of these biographical vignettes reflects typical patterns for women psychologists who came of age in the 1920s and 1930s. While sex discrimination operated blatantly at certain institutions, it was more subtle at such educational havens as Columbia Teachers College where women were consistently channeled into clinical fields. Whatever their intellectual promise and graduate training, few women could expect to find regular academic employment; those who did usually remained in low-level positions for the greater part of their careers. Most women of this generation outwardly accepted their limited professional horizons and strove to work within the discipline, despite its biased reward system.

After its formation in the wake of Pearl Harbor, the National Council of Women Psychologists began to organize on behalf of its thousand or so potential constituents. Over the next few months the Organizing Committee was actively engaged in recruiting a national membership, nominating officers, drafting a constitution, soliciting local chapters, and generally planning an agenda for the new society. But the spontaneous feminist impulse that had prompted the group's founding was soon tempered by the realities of the status hierarchy of the psychology community. Men's peacetime domination of the profession was significantly enhanced in wartime through their virtual monopoly of important advisory and supervisory posts in civilian as well as military agencies. Adopting a conciliatory strategy, the women leaders stated that "this organization should not be militant-suffragist in tone . . . [and] without stress on the fact that it is a one-sex group."[17]

This policy was implemented, at least in part, with the selection of Florence Goodenough (1886–1959) as the first president of the NCWP. Goodenough, a respected child psychologist at the University of Minnesota and a protégé of Lewis Terman, was a zealous worker who had followed the "Madame Curie strategy" of overcompensation in her career.[18] Fortunate to obtain academic employment, she not only became an excellent teacher but also pursued an extensive program of re-

16 Ibid.
17 G. C. Schwesinger to H. E. O'Shea, 12 November 1941; AIB/NCWP.
18 That is, through overachievement she competed successfully in a man's world, following a strategy identified by Margaret Rossiter as one of "quiet but deliberate overqualification, personal modesty, strong self-discipline, and infinite stoicism." Rossiter, *Women Scientists in America,* pp. 158–159, quote on p. 248.

search, concentrating on clinical methods and techniques. Her widely used Draw-A-Man Test, and her unique position as the only active female psychologist rating a "star" in *American Men of Science,* contributed to her formidable scientific reputation. Goodenough's election as president of the NCWP served a dual purpose: her high professional visibility and outstanding reputation would lend the group scientific legitimacy, while her accommodating manner and nonmilitant views made her a noncontroversial choice.[19]

There is little evidence that Goodenough was particularly interested in the NCWP and its activities – indeed, she even expressed hope for its "early demise" because she thought psychologists should not be differentiated professionally on the basis of gender. Goodenough's greatest concern was that the organization might be perceived as "advancing the cause of WOMAN, come hell or high water." Although she was willing to lend her name to an action group that would sponsor psychological services in neglected areas, she was not interested in promoting women's issues per se. Goodenough's high level of professional achievement may have shaped her uncompromising stance toward claims of sex discrimination in psychology. Eminently successful herself, she was unwilling to accept or incapable of recognizing a gender bias, often stating, "I am a psychologist, not a *woman* psychologist."[20]

Not all NCWP members shared Goodenough's attitude regarding gender discrimination. Many women psychologists in the early 1940s were beginning to recognize the subtle as well as blatant discriminatory practices operating in academic appointments and professional life more generally. The pointed exclusion of women during wartime, however, served to transform their isolated grumbling into a concerted call for feminist action. Unlike Goodenough and her kind, this core group of activists desired an organization that would represent women's causes.

The differences of opinion among these women are evident in their discussions about the goals and agenda of the NCWP. Although the purported aim of the new organization was to offer the *wartime* services of women psychologists, particularly to local communities, some of the members sought even wider scope for their frustrated energies. Thus, when the NCWP constitution and bylaws were subsequently drafted, fully 20% of the women refused to accept a clause calling for the group's disbandment after the war.[21] Their concerns transcended the immediate problems of mobilizing women psychologists and addressed the wider issue of their role within the scientific community. This action suggested that, for a significant minority at least, the "woman problem" was not simply a product of extraordinary wartime conditions, but a continuing professional issue.

19 A. I. Bryan, personal communication, 18 January 1984; T. H. Wolf, "Goodenough, Florence Laura," *Notable American Women,* 1980, 4:284–286.
20 Stevens and Gardner, *The Women of Psychology,* vol. 1, p. 195; F. L. Goodenough to Myrtle [McGraw], no date; AIB/NCWP.
21 Armstrong, "National Council."

Despite the ambivalence concerning roles and functions, the NCWP quickly gained prominence as a national organization. By the middle of 1942 the group had approximately 234 female members, all with doctorates, and four local chapters in New York, Rockland County (New York), Philadelphia, and Boston. Its legitimacy was acknowledged by an invitation to send a representative to the powerful Emergency Committee in Psychology.[22]

The NCWP clearly posed a challenge to the predominantly male experimental establishment. If not exactly threatened by the existence of the group, male psychologists were certainly made uncomfortable. While it would be overstating the case to argue that a concerted strategy was developed to counter the women's attack, the unofficial diplomacy that did evolve was aimed at deflating the issue. As it turned out, one of the most effective avenues for expressing the male viewpoint was the Subcommittee on the Services of Women Psychologists.

The Subcommittee on the Services of Women Psychologists

The Subcommittee on the Services of Women Psychologists was headed by Ruth S. Tolman (1893–1957). Married to prominent Caltech physicist Richard C. Tolman, she worked as a psychological examiner for the Los Angeles County Probation Department prior to the war. Like those of many other female psychologists, Tolman's career evolved in slow stages. After receiving a baccalaureate from the University of California in 1917, she spent the next fifteen years teaching psychology in a succession of small colleges in the Los Angeles area. In middle age she returned to Berkeley for graduate work and earned her Ph.D. in 1937 in the department headed by her brother-in-law, psychologist Edward C. Tolman. She moved to Washington, D.C., in 1941 with her husband, who was vice chairman of the powerful National Defense Research Committee, and subsequently obtained employment with social psychologist Rensis Likert in the Department of Agriculture's Division of Program Surveys. Her Washington location and, more important, her moderate stance made Tolman a natural choice to head the subcommittee.[23] Chauncey M. Louttit, who offered Tolman the chair, confirmed this impression, saying "she has no personal axe [*sic*] to grind nor is she neurotically concerned over the supposed discrimination."[24]

The other members of the subcommittee, purportedly chosen with a view to broad regional representation, included seven women and two men.[25] The group was far from radical. The majority of the female members, with the exception of Alice Bryan and Harriet O'Shea, who were both active members of the NCWP,

22 Armstrong, "National Council"; Schwesinger, "Wartime Organizational Activities."
23 Alice I. Bryan, personal communication, 18 January 1984.
24 C. M. Louttit to K. M. Dallenbach, 2 December 1941; CML/8/142.
25 Ibid.

were conservative psychologists who had openly voiced their reservations about a separate women's organization. The two men, C. M. Louttit and Steuart Henderson Britt, were driving forces behind the Subcommittee on the Listing of Personnel in Psychology and later the Office of Psychological Personnel, both of which had been the focus of contention in debates regarding the exclusion of women from mobilization efforts. When the subcommittee met for the first time in January 1942, almost two months after its establishment, its main order of business was to obtain details about the agenda and activities of the NCWP and to determine what relationship would exist between the two groups.[26] Since the NCWP was conceived as an action group, the subcommittee, to avoid duplicating efforts, regarded itself as a fact-finding and advisory organization – advisory especially with respect to the NCWP.[27] In many ways it would act as a watchdog group, and when necessary, interject a cautionary note into the proceedings of the more radical council.

The tensions generated by the charges of sex discrimination and by the formation of a national women's organization were readily apparent in discussions at the January meeting. Louttit, for instance, immediately sought to defuse the issue by emphasizing the local nature of the NCWP's activities rather than its status as a single-sex group. Community work, he noted, could "be the implementation of a lot of things if it can get organized with the proper local representation." Leaving no doubt as to his definition of "proper," however, Louttit also counseled the NCWP to "welcome or even ask for the help of men in the local groups." At every turn, the women were subtly discouraged from pursuing collective action. Even Ross G. Harrison, chairman of the National Research Council, expressed his disapproval of a separate women's group during his brief appearance at the meeting. Asked if women physicians were doing anything, he took the opportunity to admonish them: "I think not, except as doctors. I confess that when I first heard of this committee I was somewhat astonished, because I had thought of psychologists, both men and women, in these days."[28]

The male committee members were particularly anxious to deflect accusations of gender bias. Having allowed the women's grievances to escalate to the point of militancy, male psychologists were now forced, at least on a rhetorical level, to confront the issue. Although he was not a member of the subcommittee, Karl Dallenbach, chairman of the Emergency Committee, also attended the meeting and boldly raised the subject. Defensive about attacks on the committee, he adamantly "insisted that there has been no discrimination against women and that the committee is dealing with psychologists regardless of whether they be men or women." Implying that such charges had grown out of ignorance, he announced plans for a

26 R. S. Tolman to A. I. Bryan, 22 October 1941; AIB/NCWP.
27 Minutes of the Subcommittee on the Services of Women Psychologists, 10 January 1942; CML/8/142.
28 Ibid.

circular letter to explain the functions of the National Research Council and its Emergency Committee in Psychology. Dallenbach saw no need for a national organization to deal with women's issues. Although he tried to be magnanimous, he was not above hinting at the ominous consequences that would result from segregated activities:

> If the women want it, let them have it. God bless them! But I think it is too bad, because I feel that there is no restriction on a woman anywhere in the profession. I think you are raising a sex difference that is bound to have reverberations. Once you set up the difference, it will have to be accepted.[29]

When asked to address the group, the remaining male subcommittee member, Steuart Britt, also chose to speak on sexism. He had just come from a meeting considering wartime requirements for manpower in which, he pointedly added, "the term 'manpower' was used in the generic sense." His report specifically criticized women psychologists for their intemperate claims of discrimination:

> During the last few weeks, since December 7th, the work of our Subcommittee on the Listing of Personnel in Psychology has increased enormously, so that hundreds of psychologists have written in, asking for advice. Most of these have been men and they have said, "What can I do?" We are not able to tell them specifically, but when we say the same thing to women, some of them claim that they are being discriminated against. . . . I have not seen any discrimination, except in the military situation, where women could not fit into the picture. . . . If we can work on our jobs during the emergency, I think it will have to be as *psychologists*. . . . I think sometimes women defeat their own purposes by speaking of themselves as women, instead of as scientists.[30]

Britt had introduced a new dimension into the debate. Rather than the blustering denials of his colleagues, he invoked professional ideology as a means of supporting the status quo. By appealing to the women's identity as scientists, he was skillfully pressuring them to rise above divisive polemics.

Reinforcement for strict adherence to these professional norms came from the women themselves. Hence the NCWP's president, Florence Goodenough, berated her colleagues:

> It seems likely that opportunities will not be lacking for women psychologists who take their profession seriously, who are willing to compete with men on an equal basis without demanding special consideration, and who will accept the fact that no amount of faithful work on problems of little importance can compensate for lack of major scientific contribution. Women must cease to rationalize about lack of professional opportunity and demon-

---

29 Ibid.   30 Ibid. (emphasis in original).

strate their competence by actual achievement. Opportunities will expand for those who exert the necessary propulsive force.[31]

Like Goodenough, many women psychologists were willing to believe that, in the nonpolitical and unbiased world of science, merit would be the only measure by which both women and men could advance. This idealistic doctrine consequently was a powerful mechanism for maintaining the status quo and limiting agitation to a few radical women.

It is difficult to gauge what impact the men's words had on the women, or what indirect effect they had on the policies of the NCWP. At the time, however, Theodora Abel agreed that, with so much work waiting to be done, it was "unfortunate to spend any time in [*sic*] the matter of discrimination," and Helen Peak conceded that "it might be better . . . [to] includ[e] men for action organization." Even Alice Bryan, the de facto spokesperson for the NCWP, kept reiterating that it was solely an emergency organization that would disappear after the war.[32]

Given the conservative leanings of most of the female members and the active discouragement of the male members, it is hardly surprising that the subcommittee accomplished little in the way of improving the status of women psychologists. The subcommittee served mainly as an information clearinghouse, collating data on work being done, investigating areas in which women psychologists could potentially be utilized, and appointing special committees to develop war-related psychological materials. In this they relied heavily on the assistance of the NCWP. By August 1943, feeling that they had discharged their duty, the subcommittee recommended its own disbandment.[33]

Another potential venue to address gender politics was the Society for the Psychological Study of Social Issues (SPSSI). Even that organization, with its reputation as the most activist group in mainstream psychology, failed to recognize gender equity as a contemporary issue in need of social reform. Not until late 1943, after the repeated solicitations of its women members, did the organization respond by forming a Committee on Roles of Men and Women in Postwar Society. This represented SPSSI's first foray into the area of gender issues; its previous efforts had focused on social problems such as labor unrest and racial prejudice.

31 F. L. Goodenough, "Expanding Opportunities for Women Psychologists in the Post-War Period of Civil and Military Reorganization," *Psychological Bulletin,* 1944, *41:*706–712, quoted on p. 712.

32 Minutes of the Subcommittee on the Services of Women Psychologists, 10 January 1942.

33 Karl M. Dallenbach, "The Emergency Committee in Psychology, National Research Council," *American Journal of Psychology,* 1946, *59:*496–582; Ruth S. Tolman, "Wartime Organizational Activities of Women Psychologists: I. Subcommittee of the Emergency Committee on the Services of Women Psychologists," *Journal of Consulting Psychology,* 1943, *7:*296–297.

The activities of this committee were coordinated with the NCWP's Committee on Postwar Planning for Women. Under the leadership of Georgene Seward, the nucleus of the group consisted of Catherine Patrick, Elizabeth Duffy, and Gertrude Hildreth.[34] Despite Seward's ambitious plans and the committee's joint sponsorship, its visibility remained low. Gordon Allport, president of the SPSSI, admitted to having no idea of the group's objectives.[35] Seward continued to formulate the woman problem in scientific research terms. Convinced that the issue was "much broader and deeper than the old 'women's rights,' or jobs for women," its solution, she argued, would "rest upon changing certain entrenched social attitudes."[36] Besides contributing a review of biological and cultural determinants of sex roles and behavior, Seward outlined desirable changes in postwar sex roles, including the restoration of the father as a functional member of the family and an educational and cultural "renaissance" for women in society.[37] This article, however, was the sole tangible product of the committee. The committee disappeared without a trace after 1944, although Seward continued her own work on gender issues.[38] Even in a liberal and progressive association like the SPSSI, the climate was unconducive to the promotion of reforms affecting women.

The National Council of Women Psychologists thus became the only group devoted to the professional advancement of women psychologists throughout the war years. As it turned out, however, the projects promoted by the group fell short of many of the original aspirations. By electing to concentrate on civilian work, the women had circumscribed their professional sphere, and found themselves relegated to the margins of war preparation. Nonetheless they made many valuable contributions. Cooperating with local agencies, the women coordinated their efforts in a variety of areas, including psychological testing for local selective service boards, provision of child care for working parents, training of nursery school workers, evaluation and placement of volunteers in civilian defense agencies, occupational testing, and the selection of women officer candidates. The NCWP also provided lecturers to interested lay audiences and developed channels of commu-

34 G. H. Seward to G. W. Allport, 26 November 1943; GWA/Correspondence 1930–45 XYZ/SPSSI Misc. 1944–45.
35 G. W. Allport to T. M. Newcomb, 30 November 1943; GWA/Correspondence 1930–45 XYZ/SPSSI Misc. 1944–45.
36 G. H. Seward to G. W. Allport, 2 April 1944; GWA/Correspondence 1930–45 XYZ/SPSSI Misc. 1944–45.
37 G. H. Seward, "Sex Roles in Postwar Planning," *Journal of Social Psychology,* 1944, *19:*163–185.
38 G. H. Seward, *Sex and the Social Order* (New York: McGraw-Hill, 1946); G. H. Seward and K. B. Clark, "Race, Sex, and Democratic Living," in G. Murphy, ed., *Human Nature and Enduring Peace* (Boston: Houghton Mifflin, 1945); Stevens and Gardner, *Women of Psychology,* vol. 2, pp. 139–143.

nication as well as resources for those on the home front. For instance, the NCWP Committee on Publications issued numerous, widely circulated course guides and outlines for public consumption. Like most NCWP projects, these educational materials were aimed at solving community problems, such as training discussion group leaders, countering emotional strain among schoolteachers, encouraging babies to eat normally under wartime conditions, and giving "Psychological First Aid" to civilians.[39]

On the other hand, various areas of endeavor that might have expanded the women's professional sphere were neglected. Because they preferred a simple, flexible organization that could act quickly as needed, psychological research projects were, for the most part, tabled. For similar reasons the NCWP leaders decided not to protest the continued exclusion of women from War Department work. But more consequential was the NCWP's inability to place women in jobs temporarily vacated by male academics and clinicians. As growing numbers of male psychologists entered military service, women eagerly anticipated stepping into positions and roles previously denied them. Unfortunately, this major priority became subordinate in the hectic schedule of civilian service. So, although women found an important niche in wartime activities, as an organized group their expertise still remained restricted to their traditional professional sphere.

Individual Responses of Female Psychologists

Collective action, however, was just one avenue explored by women psychologists in their attempt to mobilize for war. On an individual basis they continued to confront the status quo. Working independently, some women tried to change the system through timely research projects and direct action.

Georgene Seward, for example, used her research to buttress arguments against gender-role stereotyping and sex discrimination. Her scientific work on sex roles and behavior followed an established tradition of feminist scholarship, especially in the field of sex differences pioneered by Helen Thompson Woolley and her former Columbia professor, Leta Stetter Hollingworth.[40] Seward's feminist concerns had earlier led to a series of studies on the female menstrual cycle and its effects on performance. With the influx of female workers into the labor force during the

39 Armstrong, "National Council"; Schwesinger, "Wartime Organizational Activities"; Ruth S. Tolman, "Psychological Services in the War," *Women's Work and Education,* 1943, *14*(5):1–4; Tolman, "Some Work of Women Psychologists in the War," *Journal of Consulting Psychology,* 1943, *7:*127–131.

40 See Elizabeth Scarborough and Laurel Furumoto, *Untold Lives: The First Generation of American Women Psychologists* (New York: Columbia University Press, 1987); Rosalind Rosenberg, *Beyond Separate Spheres: Intellectual Roots of Modern Feminism* (New Haven, Conn.: Yale University Press, 1982).

war, substantial concern was expressed over their fitness to perform men's work, particularly during menstruation. Defending the appropriateness of women filling such occupational roles, Seward made an exhaustive survey of the scientific evidence and came to the conclusion that there was little demonstrable relation between menstruation and work efficiency. The effects that had been observed, she argued, were based on cultural stereotypes of female behavior, not on physiological or psychological evidence.[41]

Other female psychologists, preferring more direct participation, daringly challenged and entered what was perhaps the most traditionally male institution in American society – the military. One of the first to do so was Mildred Mitchell, by then an experienced combatant in gender politics. In 1942 Mitchell, an employee of the Minnesota State Bureau of Psychological Services, took the first Women's Army Corps examination offered in the state. Upon discovering that she would not be called up immediately, she turned instead to the navy. Mitchell was allowed to enlist, but not without prejudicial treatment; whereas she was commissioned as a lieutenant, her former male student colleagues at Harvard and Yale were commissioned as lieutenant commanders. Mitchell's wartime work in clinical and personnel psychology in various navy departments made her part of an extensive and increasingly significant network. After the war she was able to transfer her experience into higher-status positions as a senior psychologist in a number of Veterans Administration centers. Her postwar career clearly highlighted the importance of military and government service roles to the professional development of psychology.[42]

Mitchell was one of the fortunate few. By the end of the war in 1945, less than forty women psychologists had served in the U.S. armed forces, compared to over a thousand of their male counterparts. Of those women, only twenty were employed as military psychologists.[43] However, female psychologists did find a niche for their professional expertise in war service – one confined largely to the home front, in civilian volunteer work.

Debate over the "Woman Problem"

Despite many trials and tribulations, female psychologists prevailed in drawing attention to their professional role and status. As the war progressed, the behind-the-scenes dialogue evolved into a public debate spearheaded by the unlikely duo of Alice Bryan and Edwin Boring.

41 G. H. Seward, "The Female Sex Rhythm," *Psychological Bulletin,* 1934, *31:*153–192; idem, "Psychological Effects of the Menstrual Cycle on Women Workers," *Psychological Bulletin,* 1944, *41:*90–102.
42 Mildred Mitchell, "Autobiography."
43 T. G. Andrews and M. Dreese, "Military Utilization of Psychologists during World War II," *American Psychologist,* 1948, *3:*533–538.

Bryan had achieved striking personal success early in the emergency. Her reputation as a team worker and her network of professional contacts combined to make her one of the most highly visible women in psychology's professional affairs. Not only had she served as executive secretary of the American Association for Applied Psychology since 1940, and as a member of the National Council of Women Psychologists governing board and the Subcommittee on Services of Women Psychologists, but she also became the NCWP's elected representative to the Emergency Committee in Psychology. In 1942 Robert Yerkes invited her to serve on the ECP Subcommittee on Survey and Planning that orchestrated the reformation of the American Psychological Association. She was the only woman in this powerful group, which included Boring among its members.

Boring and others had been piqued by Bryan's repeated assertion that women did not have proportionate representation in APA offices. After a careful examination of membership data, however, Boring admitted to Bryan that she was correct, contrary to the conception shared by Steuart Britt and Donald Marquis, both of the Office of Psychological Personnel, and him.[44] Although it would probably be stretching things to suggest that Boring left his past sexist behavior completely behind, on an *individual* basis at least, he seemed to have grown more sensitive to the plight of female professionals than he had earlier demonstrated in the case of Mildred Mitchell. In the intervening years Boring had seen his wife, who had a Ph.D. in psychology, subordinate her career to his, and was deeply affected when his older sister Alice, a teacher, returned to the United States from enemy Japan with her life's work destroyed and no prospects for the future.[45] Yet Boring was also psychology's self-appointed gadfly, being extraordinarily concerned with the discipline's public image. Few professional or scientific issues escaped his enthusiastic consideration. His interest fully aroused, Boring asked Bryan to collaborate on an empirical study of the activities and status of women psychologists. Their research resulted in three key articles published between 1944 and 1947.[46]

Although Bryan and Boring agreed that "something ought . . . to be done about women in American psychology," they approached the problem from opposing points of view.[47] In his autobiography Boring confessed that

I initiated that [collaboration with Alice Bryan] because she was a feminist who saw women as denied their professional rights, and I was on the other

44 E. G. Boring to A. I. Bryan, 10 January 1944; AIB/WIAP.
45 E. G. Boring to A. I. Bryan, 8 December 1943; EGB/Bryan 1942–43.
46 Alice I. Bryan and Edwin G. Boring, "Women in American Psychology: Prolegomenon," *Psychological Bulletin,* 1944, *41:*447–454; idem, "Women in American Psychology: Statistics from the OPP Questionnaire," *American Psychologist,* 1946, *1:*71–79; idem, "Women in American Psychology: Factors Affecting Their Professional Careers," *American Psychologist,* 1947, *2:*3–20.
47 Bryan and Boring, "Women in American Psychology: Prolegomenon," p. 447.

side thinking that women themselves for both biological and cultural reasons determined most of the conditions about which she complained. I thought
that joint study by the two of us was the way to cancel out prejudices and
leave the truth revealed.[48]

Given Boring's stated bias, it is not surprising that he saw part of his role in
the partnership to "tone down" Bryan's feminist vehemence into "humorous description."[49]

According to Boring, he and Bryan made a very good team – she had "inde-
fatigable drive" for the topic and he felt he could effectively moderate her position.[50] Drawing on data obtained from a questionnaire sent by the Office of Psychological Personnel to more than 4,500 psychologists, they found, among other
things, that full-time employment of women psychologists did not increase during
the war, and that unemployment actually increased. Contrary to popular opinion,
women did not replace men in teaching positions: although there were some four
hundred fewer male university and college teachers in 1944 than in 1940, the number of female professors remained practically unchanged. Furthermore, while male
Ph.D.s retained a median salary advantage of 20% over female Ph.D.s during the
period, the salary gap between male and female non-Ph.D.s increased. In concluding their second article, Bryan and Boring suggested that complaints of sex
discrimination were based more on "monetary compensation than upon the nature
of the jobs in which they find themselves" and claimed that proper pay would dis-
solve gender prejudice. They sidestepped the issue of why certain jobs were better paid than others in the first place and avoided dealing with how women were
tracked into them by citing gender differences in occupational interests that were
formed by tradition and culture, not by the psychology community per se.[51]

The divergent viewpoints of Bryan and Boring, never fully acknowledged in
their articles, help to account for the inconclusive and dispassionate tone of their
analysis. In private, at least, the partnership was increasingly fraught with tension.
Just three weeks after writing that their joint venture "had worked splendidly" and
that they were ready to make a definite statement regarding the status of women
in the profession, Boring confessed that he was encountering criticism from career
women, including Bryan.[52] The controversy was precipitated by Boring's strongly
worded and widely circulated draft of the "woman" paper, known as the "purple
copy" because of the color of its mimeograph ink.[53] Female psychologists accused

48 Edwin G. Boring, *Psychologist at Large* (New York: Basic Books, 1961), p. 72.
49 E. G. Boring to R. M. Elliott, 22 March 1944: AIB/WIAP.
50 Ibid.
51 Bryan and Boring, "Women in American Psychology: Statistics."
52 E. G. Boring to R. R. Sears, 26 October 1946; Boring to Sears, 18 November 1946;
    EGB/Sears 1946.
53 E. G. Boring, "The Woman–Man Argument," mimeograph, 1946; RRS/2.

Boring of only presenting material that would support his conservative theories while ignoring opposing evidence. He, in turn, charged them with emotionalism and lack of objectivity.[54] The controversy split along gender lines, and even sympathetic male colleagues like Robert Sears sided with Boring:

> Having been married to an active professional wife for fifteen years, and having been thoroughly sympathetic with women's problems during that time, and during the last four years having been director of an organization [Iowa Child Welfare Research Station] which is heavily loaded with women on the professional and student staff, I still think you have stated the situation very soundly.[55]

The outcome of this conflict was that Boring and Bryan edited the paper and toned down its scientistic ideology. Surprised at the compromise, Dael Wolfle, editor of the *American Psychologist,* apparently called Boring to ask if Bryan was "doublecrossing" him.[56]

Boring had invoked a scientific rationale based on cultural and biological determinism to explain and thereby legitimate the lower professional status of women. He and other male psychologists were unwilling to view the "woman problem" in terms of discrimination. They refused to take any responsibility, either individually or collectively, because they vigorously and persistently denied that discrimination even existed. Instead, they saw the unequal representation of women in high-status positions as a function of the relative lack of prestige of professional work relative to scientific work. As status is dependent upon prestige, Boring argued, and for biological and cultural reasons women do not have the qualifications for gaining prestige, then, in a meritocratic system such as science, they will remain in lower-level positions.[57]

Boring later acknowledged that his plan – that "by intimate discussion" he and Alice Bryan should "come into agreement" regarding women's problems – had failed. Both had remained steadfast in their values and positions, with the result that their joint papers lacked decisive explanations and conclusions. Although they did convey the magnitude and complexity of the problem, their fundamental disagreement forced them to "compromise or retreat to description of fact" rather than

54 Boring to Sears, 18 November 1946; EGB/Sears 1946.
55 R. R. Sears to E. G. Boring, 9 December 1946; EGB/Sears 1946. See also C. W. Sherif, "Bias in Psychology," in J. A. Sherman and E. T. Beck, eds., *The Prism of Sex: Essays in the Sociology of Knowledge* (Madison: University of Wisconsin Press, 1979), pp. 93–133.
56 E. G. Boring to R. R. Sears, 12 December 1946; EGB/Sears 1946.
57 Boring, "Woman–Man Argument"; Boring, "The Woman Problem," *American Psychologist,* 1951, *6*:679–682; E. G. Boring to M. B. Mitchell, 6 August 1951; EGB/Me-Mn 1951–52; F. A. Handrick, "Women in American Psychology: Publications," *American Psychologist,* 1948, *3*:541–542.

to propose any specific solutions or recommendations.[58] Frustrated, Boring was "still crazy to do something positive on this issue," especially in light of the emotion it had generated.[59]

Conclusion

The publication of the third article by Bryan and Boring in the *American Psychologist* in early 1947 marked the end of a period of debate over the role and status of women psychologists. The war had perhaps heightened the consciousness of American psychologists about gender issues, but it contributed little toward enhancing the position of women in the profession. The National Council of Women Psychologists also retreated from its coordinating role. Although a number of feminist psychologists pressed for its continuance, in 1947 the NCWP became the International Council of Psychologists. Completely abandoning its original purpose, the organization sought to "further international understanding by promoting intercultural relations [through the] practical applications of psychology."[60]

As a short-lived collective response to gender discrimination in psychology, the National Council of Women Psychologists presents certain contradictions. Why, for instance, did an organization committed to expanding the horizons of female psychologists choose to dedicate itself to what was obviously "women's work"? Although it is difficult to second-guess the motives of the NCWP organizers, several interpretations can be placed on their deliberately circumscribed role. A harsh critic could argue that, rather than disclosing any feminist agenda, the group's policies betray a defensive mentality. With men entering applied fields through government and military employment, the women may have simply been trying to protect their claims to such "feminine turf."[61] A generous critic might suggest that the NCWP had cleverly restricted its program to neglected and nonthreatening activities. Once established in wartime work – their foot in the door – the women could more easily infiltrate other areas. A third interpretation, and perhaps the most likely, strikes a balance between the previous two. The creation of the NCWP represented a gentlewomanly protest against being ignored as professionals. As such, it formed the reluctant and ambivalent core of an effort to reform psychology along

58 Boring, *Psychologist at Large*, p. 185.

59 Boring to Sears, 12 December 1946; EGB/Sears 1946.

60 L. G. Portenier, ed., *The International Council of Psychologists, Inc.: The First Quarter-Century, 1942–1967* (Greeley, Colo.: ICP, 1967), p. 23; E. M. Carrington, "History and Purposes of the International Council of Women Psychologists," *American Psychologist,* 1952, 7:100.

61 L. J. Finison and L. Furumoto, "An Historical Perspective on Psychology, Social Action and Women's Rights," unpublished manuscript, 1978.

feminist lines. With the onset of the emergency situation, the women asked only that they be treated as psychologists. But faced with the prospect of exclusion from significant war work, and after being treated discourteously, they were swept into an unprecedented independent action that had wider feminist implications.

Looking back, Alice Bryan noted that,

> had the dominant male members [of the AAAP and APA] been sufficiently skillful as psychologists, they would have kept the women informed about progress toward organizing the profession to aid in the war effort, and assured them that they would be participants. Instead, the women were put off and told to be "good girls" and "to be quiet." . . . If the women had been treated with proper deference and quickly given some low status appointments on the subcommittees of the EC [Emergency Committee in Psychology], the National Council of Women Psychologists would never have been organized.[62]

The women's subsequent actions reflect a complex mixture of consciousness raising and conservative reaction. The fluid wartime environment had provided women with an opportunity to vocalize concerns about gender discrimination that had long gone unstated. Their ambivalence about this role, however, was expressed in limiting the scope of their work and in electing noncontroversial representatives like Florence Goodenough and Ruth Tolman. Tolman, in particular, may not have had a realistic understanding of the discrimination that was operating against many of her female colleagues. Although she had faced the indignity of a succession of low-paying positions in small colleges before achieving career stability in middle age, she had never been at the margins of the scientific community. As the wife of physicist Richard Tolman and the sister-in-law of psychologist Edward Tolman, she was in close if informal contact with members of the scientific elite. Her unique social position may have made it difficult for her to perceive the hindrances faced by her colleagues who traveled in less exalted circles. Tolman might in fact have been an exception to Boring's rule that marriage hurt the careers of professional women. As chairman of the Subcommittee on the Services of Women Psychologists, she worked conscientiously to promote projects for women. Nonetheless, habits of deference were hard to overcome, and after the disbandment of the subcommittee she wrote to Louttit: "I feel that your male influence probably saved us from many a folly!"[63]

The ambivalent attitudes expressed by many female psychologists had been shaped by the immediate context of the war as well as by more general social and

---

62 A. I. Bryan, personal communication, 22 December 1984. See also Bryan, "A Participant's View of the National Council of Women Psychologists: Comment on Capshew and Laszlo," *Journal of Social Issues,* 1986, *42*(1):181–184.

63 R. S. Tolman to C. M. Louttit, 3 September 1943; CML/8/142.

cultural forces. Although the wartime environment did allow American women access to many traditionally male occupations, it was seen as a temporary necessity. Awareness of women's rights was limited, and women were expected to relinquish their places to the returning servicemen. In this unresponsive atmosphere, female psychologists found little support or encouragement for a feminist agenda. Perhaps even more than this negative climate, an entrenched ideology of scientific professionalism contributed to the ambivalence of female psychologists. Ironically, the low goals set by women's groups during the war may be taken as an indicator of their high degree of professionalization. Thoroughly socialized in an elitist scientific community, the majority of female psychologists found it more palatable to attribute their low status to low achievement rather than to any conscious discrimination. Because all "objective data" pointed to women's preference for applied work while the scientific value system accorded higher prestige to theoretical and experimental research, they may have conceded a certain justification for their lower status.

Female psychologists had yet to develop an identity for themselves as a group whose collective experiences would reveal a pattern of professional discrimination. They would have had no difficulty appreciating Ruth Tolman's comment that she found "it hard to abstract 'being a woman' from being a particular woman and tend to hold responsible my particular idiosyncrasies rather than my sex for the arrangements of my life."[64] In this context, Boring's treatment of the "woman problem," with its conscientious invocation of a scientific protocol, merely served to remind psychologists of what they had already internalized.

64 R. S. Tolman to E. G. Boring, 25 January 1946; EGB/58.

Interlude II

A History of Experimental Psychology was not the comprehensive work that Edwin Boring had envisioned. His original plan had been "to start with the men and the schools as an introduction and then to trace the history of experimentation and thought in the fields of sensation, perception, feeling, emotion, learning, memory, attention, action and thought." Despite its title, the 1929 book contained only the introductory portion of the larger project, and Boring belatedly realized that he "had not yet got to the experimentation in experimental psychology." Throughout the 1930s he worked on the conceptual and technical history of research in sensation and perception, which he considered the core of scientific psychology.[1]

In addition to his historical research, Boring tried his hand at textbook writing. Concerned that introductory psychology books did not emphasize strongly enough the findings of experimental research, he collaborated with two longtime professional allies to produce *Psychology: A Factual Textbook,* published in 1935. Boring's coauthors were Herbert S. Langfeld, his predecessor at Harvard and then head of the psychology department at Princeton, and Harry P. Weld, a professor at Cornell whom he had known since before the First World War. All three men shared a strong positive bias toward the experimental tradition in psychology.

Seeking to produce "an authoritative text" that avoided wrangling over theoretical perspectives, the trio wanted the book "to be science but to have no special point of view."[2] Their approach resulted in a text that was heavily slanted toward sensation, perception, the nervous system, and other standard topics.[3] The market for undergraduate textbooks was changing, however, as interest in the problems of

1 Edwin G. Boring, *Sensation and Perception in the History of Psychology* (New York: Appleton-Century-Crofts, 1942), p. vii.
2 Edwin G. Boring, *Psychologist at Large* (New York: Basic Books, 1961), p. 57.
3 Edwin Garrigues Boring, Herbert Sidney Langfeld, and Harry Porter Weld, *Psychology: A Factual Textbook* (New York: John Wiley & Sons, 1935).

human adjustment was reflected in the rise of social, clinical, and other forms of applied psychology. In this environment, and in response to the suggestions of some of their colleagues, Boring, Langfeld, and Weld revised their textbook a few years later and added more material on such subjects. The result was *Introduction to Psychology,* published in 1939.[4] Although most users welcomed the changes, according to Boring, some "staunch conservatives" did not and continued to use the earlier edition.[5]

In 1942 Boring finally published the second volume of his planned historical trilogy. Entitled *Sensation and Perception in the History of Experimental Psychology,* the book was dedicated to Hermann von Helmholtz. It reviewed the intellectual history of sensory science, dealing with each sense modality (vision, hearing, smell, taste, touch) in turn. Although it retained the previous volume's biographical approach, it focused more on the development of ideas. In the brief concluding chapter, "Concerning Scientific Progress," Boring sketched his views on the history and psychology of scientific thought. Seeking to explain how impersonal, objective knowledge can be generated almost "in spite of the limitations of the human mind," he introduced the notion of the Zeitgeist as a potential facilitating or inhibiting factor. Thus, intellectual progress still depended on individual thinking, but that thinking was shaped not only by the personality of the investigator but also by social attitudes and collective habits of thought beyond the control of any single individual.

Sensation and Perception in the History of Experimental Psychology proved to be Boring's last major historical treatise. After completing the book, he dated the preface 6 December 1941, calling it "the last day when pure scholarship could be undertaken with a clear conscience."[6] This trifling literary conceit, coming after Boring was already caught up in wartime mobilization efforts, was merely a means of dramatizing his sense that momentous changes were in store, for himself and for psychology.

After Pearl Harbor, Boring turned his full attention to war work. He watched his younger colleagues go off to active duty, as he had in World War I, and saw old friends such as Robert Yerkes and Karl Dallenbach find advisory posts in Washington. As his Harvard colleagues departed, Boring shouldered a heavier teaching load, but he continued to search for some project more directly relevant to mobilization for defense.

Although he was not an official member of psychology's "war cabinet" – the Emergency Committee in Psychology – Boring knew most of its members well

---

4 Edwin Garrigues Boring, Herbert Sidney Langfeld, and Harry Porter Weld, *Introduction to Psychology* (New York: John Wiley & Sons, 1939).
5 Boring, *Psychologist at Large,* p. 58.
6 Ibid., p. 60.

and followed its activities closely. During their early meetings, the group discussed the need for a textbook in military psychology, an idea that had been around since the First World War. Such a book "would show what psychology had accomplished and could offer the armed services" and might be used at the U.S. Military Academy at West Point. But the proposal was shelved because no suitable author was available.[7] Boring revived the idea and suggested that a volume be produced in the same way that he and Langfeld and Weld had composed their introductory psychology textbooks, with chapters drafted by various specialists and then edited into a unified whole. In May 1942 the Emergency Committee accepted Boring's plan and appointed him chairman of the Subcommittee on a Textbook in Military Psychology.[8]

The veteran author found the prospect of writing his sixth book "wonderful therapy" for his frustration and quickly got organized for the task. The planned book was divided into four sections (motivation and morale, training and selection, propaganda and psychological warfare, perceptual functions); editors for each section were assigned; and contributors were recruited to write individual chapters. Not surprisingly, the usual delays in academic collaboration were compounded by the national state of emergency, and material trickled in slowly. In the meantime, the thrust of the book was reconsidered. At the urging of military advisors, the plan for a college textbook aimed at West Point cadets was abandoned in favor of a popular treatise geared toward the average soldier in the field. A cheap mass-market paperback with a press run of 60,000 to 100,000 copies was proposed.

Although Boring was confident that he could change his normal academic writing style into something more accessible, he turned to a professional science journalist, Marjorie Van de Water (1900–1962), for assistance. Van de Water had covered the psychology beat for the Washington-based Science Service news organization through the 1930s and had a knack for translating esoteric research findings into readable accounts for the lay public. Her contacts, especially among applied psychologists, were extensive. Indeed, at the beginning of 1941 she had been recruited to chair the American Association for Applied Psychology's Committee on Public Relations.[9]

Van de Water joined the paperback book project in August 1942 as coeditor and the collaboration with Boring began in earnest. She countered his dominating personality with her own strong assertiveness as she became his tutor in popular science writing. From the start she helped to shape not only the style of manuscript but also the selection and emphasis of topics. Eschewing chapters organized under

7 Karl M. Dallenbach, "The Emergency Committee in Psychology, National Research Council," *American Journal of Psychology,* 1946, *59:*496–582, quote on p. 526.

8 Ibid., pp. 526–527.

9 See James H. Capshew, "Psychology on the March: American Psychologists and World War II" (Ph.D. diss., University of Pennsylvania, 1986), chap. 5.

traditional academic rubrics, she pushed for a problem-based approach that concentrated on the psychological aspects of various military situations.

As new material continued to arrive from contributing authors it became clear to both Boring and Van de Water that they would have to rewrite nearly all of it. Many authors had trouble avoiding academic jargon and pedantry; others wrote *about* the soldier rather than directing their remarks *to* him. Van de Water was impatient with contributors who were too abstract or theoretical, commenting to her coeditor: "I would like to see more facts included. I think the book should bristle with them."[10]

Although Boring was impressed by Van de Water's professional skills, he expressed some ambivalence about the project and their respective roles in the collaboration. Early on he asked her if she had ever considered using a pseudonym:

> If you were named Mandel Waters, for instance, it would help us a lot. I have been advised that no woman can be played up for the Army, but in a case like this there arises both a problem of credit and discredit. You ought to have credit as a third author [of a particular section], and it might be that other authors would also like to have someone to blame the style on. I am enthusiastic about the style, but I should hesitate over [publishing under] my name something you had written because those who know me would know that it had been ghost-written, that I never could have achieved it. But if some other name could be added to the title, then that would take all the responsibility off the original authors who furnished the ideas.

Insulted by Boring's suggestion, Van de Water retorted, "I am seriously opposed to the use of an assumed name for any purpose – especially in wartime." She was not averse to using the phrase "with the collaboration of Science Service" if he wished, as long as her contribution was properly credited in the acknowledgments.

Boring was also uncomfortable about the book's applied orientation, even though he knew that military psychology was fundamentally concerned with putting knowledge into practice. Work in nonexperimental areas such as social, clinical, and personnel psychology was obviously relevant to military interests, but Boring believed that much of the discussion of factual matters was contaminated by unwarranted value judgments. He complained to Van de Water that "these academic writers are trying to bring [in] their feeling of social consciousness and that it very frequently needs to be taken out. This is not a book in which to complain about the society as it is." Despite his qualms, however, Boring pressed on with the task.

Boring often viewed psychology in polarized and oppositional terms, and sometimes saw himself as a mediator. As *Psychology for the Fighting Man* began to take shape, he noted: "The book tends to move between two poles; one, dense but dull

10 Ibid., p. 217.

(perceptual capacities), the other, interesting but thin (social relations). The planners of this book worried whether there would be too much of one or of the other. . . . Perhaps the venture got underway only because some thought that *I* could resolve this polarity."[11]

In March 1943 Boring and Van de Water finished the book and sent the manuscript to press. The 450-page paperback appeared in July, and by the end of the summer 250,000 copies were in print. Although it garnered positive reviews and eventually sold over 400,000 copies – making it perhaps the most widely distributed psychology book ever – Boring was not sure exactly what he had wrought in producing a vade mecum for the soldier. Soon after their collaboration ended, he wrote to Van de Water, musing: "I think the Fighting Man will leave the reader with the impression that man is a mechanism, that there are laws that govern his actions, that he ought to take that point of view toward human problems in the Army, that psychology is a great thing. That is all I thought the Fighting Man was for. . . ." Hardly stopping to rest, Boring revived the idea of a West Point textbook on military psychology and began working on yet another manuscript.

Although the military psychology textbook was organized along the same lines as *Psychology for the Fighting Man*, Boring tried to make it more academic in tone. He included more material from traditional experimental areas in a self-conscious attempt to be more scientific. The resulting book, *Psychology for the Armed Services,* was published in July 1945.[12]

The volume represented a hybrid offspring of *Psychology for the Fighting Man* and the Boring, Langfeld, and Weld introductory textbooks. As such, it proved less than successful for the army's educational purposes. Although the book received some use, Boring realized that West Point "wanted tougher chewing for their cadets" and wished he had conveyed "more the odor of science, less the tang of human nature."[13]

After the war Boring continued to work through his ambivalence toward applied psychology by writing yet another textbook. Joining once again with Langfeld and Weld, he spearheaded the production of the third revision of their introduction to psychology. *Foundations of Psychology,* published in 1948, attempted to capture part of the expanding postwar textbook market with an updated version of experimentalism.[14]

As before, Boring dominated the collaboration. The book's tone and coverage

---

11 Boring to Van de Water, 6 September 1942; EGB/60/1368.

12 Edwin G. Boring, ed., *Psychology for the Armed Services* (Washington, D.C.: Infantry Journal, 1945).

13 Boring, *Psychologist at Large,* p. 63.

14 Edwin Garrigues Boring, Herbert Sidney Langfeld, and Harry Porter Weld, *Foundations of Psychology* (New York: Wiley & Sons, 1948).

reflected his own evolving sense of the proper definition of psychological science and his ongoing struggle to fit applied psychology into his disciplinary framework. The experiences of writing *Psychology for the Fighting Man* and *Psychology for the Armed Services* had left their mark. The new introductory textbook included much more material on social, clinical, and applied psychology than its 1930s predecessors. In fact, nearly half of the 1948 edition dealt with material drawn from outside traditional experimental topics, compared to only about 20% from the 1935 edition.[15] In the process of dealing with the rapid rise of applied psychology during the 1940s, Boring struggled to find a viable accommodation between his idealistic vision of experimental science and the realities of professional practice. His transition from foe to friend of professionalization was reflected literally in the pages of the five popular and introductory texts he wrote between 1935 and 1948. As Boring assimilated the lessons of war, his writing moved into the mainstream, sharing the ecumenical vision of psychology as a diverse but unified scientific profession.[16]

15 Capshew, "Psychology on the March," p. 230.
16 See Dael Wolfle, "The First Course in Psychology," *Psychological Bulletin,* 1942, *39:* 685–712; Ned Levine, Colin Worboys, and Martin Taylor, "Psychology and the 'Psychology' Textbook: A Social Demographic Study," *Human Relations,* 1973, *26:*467–478.

4

Sorting Soldiers
Psychology as Personnel Management

In the summer of 1943 a paperback book entitled *Psychology for the Fighting Man* appeared in U.S. bookstores and Allied post exchanges around the world. The pocket-sized Penguin edition, costing twenty-five cents, joined a list of books aimed at the wartime mass market, both military and civilian. Directed to "the fighting man himself," the volume adopted a conversational tone in explaining "what you should know about yourself and others."[1]

From its bright red cover to its snappy chapter titles, *Psychology for the Fighting Man* was designed to capture the attention of the average reader. The twenty short chapters were written to stand individually and could be read in any order. The topics ranged across the entire spectrum of contemporary psychology, from motivation ("Morale") to sensation and perception ("Sight as a Weapon"), from personnel selection ("The Right Soldier in the Right Job") to social and cultural psychology ("Differences among Races and Peoples").

Psychology for the Fighting Man became a bestseller, with nearly 400,000 copies in circulation by the end of the war. The book was reviewed widely upon publication. The press was positive, if somewhat bland, in its evaluations. The *New York Times Book Review* called it "a why-you-behave-as-you-do sort of a book, scientifically accurate, militarily correct and keyed to interest men of action whether colonels or corporals or anyone literate enough to read a newspaper."[2] Psychologists gave it high praise and tended to find their own prejudices confirmed in its pages. For instance, behaviorist Walter Hunter noted approvingly the heavy emphasis on human sensory and perceptual capacities and suggested that college

1 National Research Council, *Psychology for the Fighting Man* (Washington, D.C.: Infantry Journal, 1943).
2 S. T. Williamson review of *Psychology for the Fighting Man, New York Times Book Review,* 22 August 1943, p. 20.

textbook authors expand their treatment of such topics.[3] Clinical psychologist Laurence Shaffer thought the book's strong focus on personal adjustment and social relations lent support to social and applied psychology as legitimate fields.[4] Overall, psychologists lauded the book as an accurate portrayal of psychology's scientifically grounded social utility and were pleased with the public attention it drew to their work.[5]

The significance of *Psychology for the Fighting Man* went well beyond its role in popularizing military psychology. Although psychologists had been writing for general audiences for decades, this book had the imprimatur of the psychology establishment. Produced under the aegis of the National Research Council, the book listed more than fifty collaborators in a virtual who's who of academic psychology. Furthermore, the inclusion of "softer" subjects like morale or leadership alongside "harder" topics such as night vision served as a indication of how disciplinary diversity had been managed for the sake of professional unity. As the first popular science book officially sponsored by the profession, it can be viewed as a textual artifact that documents a changing ideology in psychology. Its pages capture many of the important motifs of wartime American psychology, including its utilitarian slant, its preoccupation with social rather than scientific problems, and its aggressive public relations. Furthermore, the production of *Psychology for the Fighting Man,* spearheaded by Edwin Boring, one of psychology's opinion leaders, provides an index of changing attitudes toward applied psychology.

The volume's contents revealed the reorientation of the discipline under wartime conditions. The book covered many traditional scientific subjects, including sensation, perception, and learning, but conveyed their findings in a novel fashion that made them relevant to the wartime situation. For instance, knowledge about the functioning of the human eye's rods and cones informed the practical suggestion of using red goggles to adapt the eyes to darkness before night duty. In contrast, the findings of applied psychology in areas such as personal adjustment, vocational selection, and morale required less translation for military usefulness. Chapters on psychological warfare and international relations invoked the findings of social psychology and cultural psychology to argue against racial and cultural stereotypes.

---

3 He estimated that nearly one-third of the text dealt with such areas, while college textbooks customarily devoted only 10% of their pages to them. Walter S. Hunter, "*Psychology for the Fighting Man: A Special Review,*" *Psychological Bulletin,* 1943, *40:*595–597.

4 Laurance F. Shaffer, review of *Psychology for the Fighting Man, Journal of Consulting Psychology,* 1944, *8:*50–51.

5 Other reviews included: Leonard Carmichael, *Science,* 1943, *98:*242–243; Helen Peak, *Public Opinion Quarterly,* 1944, *8:*272–275; and Paul S. Farnsworth, *American Sociological Review,* 1943, *8:*733–734.

Psychology for the Fighting Man showed how the diverse content of psychology could be brought under the umbrella of personal adjustment. It offered a unifying conceptual framework that harmonized the findings of "pure" laboratory research with the practical thrust of applied psychology. Behind the image of the usefulness of psychology portrayed in the book was the reality of the profession's actual wartime employment. The bulk of the nation's psychologists were involved in the management of military personnel. The war required the selection and training of millions of soldiers, and psychology offered practical tests and techniques to rationalize the process.

Military personnel work, including the classification, selection, and training of soldiers, was a major arena for the activities of psychologists during the war. The recruitment and deployment of millions of civilians into the armed forces was a massive job, and psychologists negotiated an important role for themselves in the military personnel system. Psychologists drew on the lessons of the First World War as well as from professional resources garnered in the interwar years to help establish the basic structure of the army personnel system, including the ubiquitous Army General Classification Test. Concentrating first on integrating the normal recruit into army life, psychologists later turned their attention to the problems of illiterate draftees and eventually to clinical work in army hospitals and rehabilitation centers.[6]

The Army Classification Program

In the summer of 1939 the United States had a standing army of approximately 174,000 men. After President Roosevelt declared a limited national emergency in September, the army was authorized to increase its strength to 227,000 troops, along with 235,000 reserves in the National Guard. Faced with the probability of large future increases, the Adjutant General's Office (AGO) had to revamp its entire personnel system, which remained little changed after demobilization for World War I. AGO officers began discussing personnel problems with Walter V.

6 This chapter concentrates on army (including army air forces) personnel work. Psychologists also made significant contributions to the navy's system; see C. M. Louttit, "History of Psychological Examining in the U.S. Navy," *U.S. Navy Medical Bulletin*, 1942, *40:*663–664; idem, "Psychological Work in the United States Navy," *Journal of Consulting Psychology*, 1941, *5:*225–227; idem, "A Study of 400 Psychologists Commissioned in the U.S. Naval Reserve," *Psychological Bulletin*, 1944, *41:*253–257; Dewey B. Stuit, ed., *Personnel Research and Test Development in the Bureau of Naval Personnel* (Princeton, N.J.: Princeton University Press, 1947); and George G. Killinger, ed., *The Psychobiological Program of the War Shipping Administration* (Stanford, Calif.: Stanford University Press, 1947). For a general overview, see Frederick B. Davis, *Utilizing Human Talent: Armed Services Selection and Classification Procedures* (Washington, D.C.: American Council on Education, 1947).

Bingham and other psychologists in late 1939.[7] They visited Chanute Field in Illinois, where psychologist Richard W. Faubion was using an adaptation of the Group Examination Alpha test from World War I to help select air corps mechanics. Within months Faubion, along with his advisors, Marion Richardson (University of Chicago) and Thomas Harrell (University of Illinois), found themselves employed by the federal government to help set up a testing program for the army.[8]

The Personnel Research Section (originally the Personnel Testing Section) was set up by Faubion in the spring of 1940. Bingham, former head of the Personnel Research Federation and a widely known applied psychologist, soon emerged as the leader of the group of consultants and was appointed chair of the AGO's Committee on Classification of Military Personnel (CCMP) by May 1940.[9] Unlike its World War I counterpart, the CCMP was an advisory group rather than an operating body. Bingham had been executive secretary of the army's Committee on Classification of Personnel in the First World War and was able to cite its accomplishments in support of his plans for the new committee's work. Particularly useful was the two-volume report, entitled *The Personnel System of the United States Army,* produced in 1919, which described the development and operation of the program. In contrast to the controversial intelligence testing program directed by Yerkes in World War I, psychologists' personnel work was less well known. But it had an important effect in generating support within the military bureaucracy for the use of psychologists and their techniques. In many ways Bingham and Yerkes recapitulated their World War I roles in World War II: Bingham accommodated his plans within the existing military structure whereas Yerkes attempted to create a distinctive institutional niche for psychology, as the disciplines of medicine and chemistry had in the Medical Corps and the Chemical Warfare Service.[10]

7 Only a year earlier the AGO had refused offers of help from veterans of World War I psychology programs, including Horace B. English and Walter Dill Scott, chair of the AGO's Committee on Classification of Personnel in World War I. Donald S. Napoli, *Architects of Adjustment* (Port Washington, N.Y.: Kennikat Press, 1981), p. 88.

8 Roy K. Davenport and Felix Kampschroer, eds., *Personnel Utilization: Classification and Assignment of Military Personnel in the Army of the United States during World War II* (unpublished manuscript, Army History Office, 1947), chap. 2; ACMH/196/4-1.4 BA.

9 Other members were: C. C. Brigham, H. E. Garrett, L. J. O'Rourke, C. L. Shartle, and L. L. Thurstone.

10 U.S. Adjutant General's Office, *The Personnel System of the United States Army,* 2 vols. (Washington, D.C., 1919). For a thorough account of World War I personnel work, see Richard T. von Mayrhauser, "The Triumph of Utility: The Forgotten Clash of American Psychologies in World War I" (Ph.D. diss., University of Chicago, 1986). On World War I intelligence testing and its aftermath, see Franz Samelson, "Putting Psychology on the Map: Ideology and Intelligence Testing," in Allan R. Buss, ed., *Psychology in Social Context* (New York: Irvington, 1979), pp. 103–168; Daniel J. Kevles, "Testing the

The CCMP worked closely with the Personnel Research Section (PRS) in preparing basic classification tools for the army, including a general aptitude test, a card for soldier personnel data, and various specialized tests. In the summer of 1940 the first version of the Army General Classification Test (AGCT) was prepared for standardization. The Soldier's Qualification Card, which had proved highly useful in the First World War, was revised, and provided with a set of punchholes along its margin to allow for mechanical sorting. But these preparations were barely underway when the Selective Service Act was passed in September 1940. A month later 16,000,000 male citizens aged 21 to 36 registered with local draft boards. As part of its effort to cope with the situation, the AGO decentralized personnel operations by reactivating personnel units on the battalion, regiment, and division levels. (Personnel responsibilities had been centralized in the AGO in 1933.) This action created new opportunities for personnel experts, including psychologists, while at the same time it made the coordination of field activities with headquarters' policy more complex and difficult, and throughout the war lines of responsibility for classification and assignment remained blurred.[11]

The Army General Classification Test

The Army General Classification Test (AGCT) was the basic personnel selection tool in the war. Prepared under pressure of impending mobilization, it was viewed as an essential sorting device. Mental testing, particularly the notion of the I.Q. (Intelligence Quotient), had seized the public imagination in the interwar years, partly as a result of the army's testing program in the First World War. Psychologists were eager to capitalize on this widespread knowledge but were also concerned about the negative connotations of testing. They tried to use this notoriety as a positive resource and attempted to shape perceptions of their proper role in personnel work.[12]

Despite questions about its scientific validity, the AGCT was pushed as a necessary component of the classification program. Psychologists were aware that it

Army's Intelligence: Psychologists and the Military in World War I," *Journal of American History,* 1968, *55:*565–581; John Carson, "Army Alpha, Army Brass, and the Search for Army Intelligence," *Isis,* 1993, *84:*278–309. See also Thomas M. Camfield, "Psychologists at War: The History of American Psychology and the First World War" (Ph.D. diss., University of Texas, 1969).

11 Davenport and Kampschroer, *Personnel Utilization,* chap. 2.

12 Bingham's comments to Yerkes in 1941 illuminate this attitude: "What we want to do is to dissociate in the minds of officers and public . . . any comparisons between Army test performance and purely developmental standards, concepts, and units." He went on to admit that "this . . . is not easy to do even within our profession." Bingham to Yerkes, 15 July 1941; RMY/95/1796.

fulfilled the needs of the army for a rational sorting mechanism, and their scientific consciences could be quieted by referring to the exigencies of war.[13]

Avoiding loaded terms such as "intelligence," "I.Q.," or even "mental," psychologists conceptualized the AGCT as a test of general learning ability or "trainability." Sidestepping the controversial issue of the relative importance of innate versus learned factors, the AGCT score was considered a measure of the combined effects of native endowment and education. The emphasis was on measuring intellectual power rather than speed, so ample time was allowed for taking the test, and the items were gradated in difficulty from very easy to fairly hard. Seeking to demystify testing procedures, the psychologists tried to make the test as practical and nonthreatening as possible. As Marion Richardson, one of its major architects, noted:

> It was decided that the test should be readily scored by hand, as well as by machines; that it ought not to have an esoteric or puzzle-like appearance, but should appeal to both officers and men as a sensible practical test so that they would take it seriously and have some confidence in the fairness and worthwhileness of the scores. This means that it must be free from ludicrous items, and items that look childish, schoolish, bookish or otherwise out of place in a test taken by mature men who may or may not have done much reading or writing since their school days.[14]

The first version, the AGCT 1a, was released in October 1940 and was standardized on a group of army and civilian conservation corps men. Version 1b followed in April 1941; six months later versions 1c and 1d were introduced and used for the remainder of the war. A revised version – the AGCT 3 – was produced for the postwar army in 1945. The test consisted of fifty items dealing with vocabulary, arithmetic, and block counting, and took forty minutes to administer. Each question had four multiple-choice answers. The raw scores, computed by subtracting one-third of the wrong answers from the number of correct answers, were converted into standard scores with a mean of 100 and a standard deviation of 20. Based on the standard scores, five army grades were established:

Army Grade	Score
I	130 and above
II	110–129
III	90–109
IV	60–89
V	below 60

13 As Marion Richardson, one of the test's architects, noted: "Early in 1940 a general classification test was prepared and adopted as a stopgap procedure, even though research conducted outside the War Department between the wars gave rise to much doubt concerning the concept of a single intelligence or general mental ability." Quoted in Davenport and Kampschroer, *Personnel Utilization,* p. 17.

14 Quoted in Davenport and Kampschroer, *Personnel Utilization,* p. 62.

The reliability of the Army General Classification Test was estimated by a variety of measures to be above .90. Its validity was good in selecting men for specialized training. Not surprisingly, AGCT scores were highly correlated with educational level as well as other tests of mental ability.[15]

The Personnel Research Section

The Personnel Research Section of the Adjutant General's Office performed the bulk of research and development in the army's testing program. Working closely with the Committee on the Classification of Military Personnel, the section assembled a large staff of military and civilian psychologists. After its start in 1940, the Personnel Research Section concentrated on devices for initial classification and specialist training selection. The AGCT and mechanical and clerical aptitude tests were among the major projects. As efforts went forward on a variety of tests, specialized working groups were formed. In response to the inauguration of the Army Specialized Training Program (ASTP) in 1943, the group moved to New York City and greatly increased its staff in order to develop achievement tests for that program. By war's end, nearly two hundred people had served in the PRS, and over eight hundred research reports had been published.[16]

Job analysis, another major component of personnel work, was undergoing simultaneous development with the testing program. By the middle of 1939 it had become obvious that serious reforms were necessary in army job rating and pay scheduling procedures, and the U.S. Employment Service was called in to help devise a rational and coordinated system. The Employment Service had launched an ambitious occupational research program in the 1930s and had just published the first *Dictionary of Occupational Titles*.[17] Psychologist Carroll Shartle was head of

15 For an overview of the development of the AGCT, see: Staff, Personnel Research Section, AGO, "The Army General Classification Test," *Psychological Bulletin*, 1945, *42:* 760–768. Technical developments are discussed in E. Donald Sisson, "The Personnel Research Program of the Adjutant General's Office of the United States Army," *Review of Educational Research*, 1948, *18:*575–614.

16 Staff, Personnel Research Section, AGO, "Personnel Research in the Army: I. Background and Organization," *Psychological Bulletin*, 1943, *40:*129–135; idem, "Personnel Research Section, The Adjutant General's Office: Development and Current Status," *Psychological Bulletin*, 1945, *42:*445–452. For a bibliography, see Sisson, "The Personnel Research Program of the Adjutant General's Office of the United States Army."

17 U.S. Employment Service, *Dictionary of Occupational Titles* (Washington, D.C.: GPO, 1939). For an overview, see Carroll L. Shartle, "The Occupational Research Program: An Example of Research Utilization," in *Case Studies in Bringing Behavioral Science into Use* (Stanford, Calif.: Stanford University, Institute for Communications Research, 1961), pp. 59–74. See also Carroll L. Shartle et al., "Ten Years of Occupational Research," *Occupations*, 1944, *22:*387–446; W. H. Stead and W. E. Masincup, *The Occupational Research Program of the United States Employment Service* (Chicago: Public Administration Service, 1941).

the Occupational Analysis Section (OAS), and its Technical Board included a number of psychologists, including Bingham and O'Rourke. Having already analyzed some 55,000 jobs in American industry, the occupational analysts were deployed to army posts in the spring of 1940 and conducted an army-wide survey. Using the *Dictionary of Occupational Titles,* civilian job classifications were converted for military use. One aspect of OAS research proved to be especially valuable: the use of job family classifications, which facilitated the transfer of workers skilled in one occupation into similar related occupations. In addition to the guidance provided by occupational classification, the army also used OAS members in test development and standardization.[18]

Army Personnel Consultants

As the army personnel system got under way, it became obvious that special classification officers would be needed in the field. Before September 1940 neither army regulations nor tables of organization provided for classification of officers or enlisted men.[19] At first, experienced soldiers were diverted into use, and the War Department staff, including Bingham, facilitated the employment of qualified reserve officers. A classification school was begun under the AGO in Arlington, Virginia, and by March 1942 had graduated nearly 1,500 classification specialists.[20] In 1942 more formal procurement procedures were begun. A three-month-long Officer Candidate School was established, eventually graduating over 1,000 specialists. To be considered for classification training, enlisted personnel needed a minimum AGCT score of 100 and experience in a relevant civilian occupation. Of the more than 5,000 enlisted graduates, 40% remained with their original units and the rest received assignments from the AGO.[21]

Psychologists viewed this expansion of the army's existing personnel system as necessary, and they aided it by serving as civilian consultants and teachers and as uniformed participants. But they felt that more highly trained psychologists were needed in army field units. As early as 1940 Bingham had suggested a study of army jobs that could be best performed by psychologists. He and Yerkes petitioned the secretary of war, Henry Stimson, in early 1941 for greater use and recognition of military psychology, perhaps in the Army Specialist Corps. The AGO, however,

---

18 Carroll L. Shartle, "New Defense Personnel Techniques," *Occupations,* 1941, *19:*403–408. The OAS was transferred to the War Manpower Commission in February 1943. Local draft boards were also supplied with the *Dictionary of Occupational Titles,* but it proved to be of little use in that context.

19 Davenport and Kampschroer, *Personnel Utilization,* p. 305.

20 Morton A. Seidenfeld, "The Adjutant General's School and the Training of Psychological Personnel for the Army," *Psychological Bulletin,* 1942, *39:*381–384.

21 Davenport and Kampschroer, *Personnel Utilization,* pp. 307–318.

did not concur with the suggestion to extend psychological work and wanted to retain psychologists in the AGO, pointing out the problems that arose from having specialists without military training which had occurred in the First World War. Instead they suggested a possible forty-seven-week course of special training from induction through Officer Candidate School.[22]

Gradually a formal role for psychologists in the field program evolved. Although most AGO officers at headquarters appreciated the value of psychological techniques in the personnel system, they were well aware of the popular conception of psychologists as "crackpots" that pervaded the army. The confusion over the proper title for psychological specialists mirrored the lack of general agreement over their proper duties. Among the terms suggested were "military psychologist," "psychological examiner," "psychometrist," "psychologist officer," "specialist in human engineering," and "classification officer." The latter title was out because the job was to be distinct from and subordinate to that of the existing classification officer. Eventually "military psychologist" and "psychological examiner" were chosen, although the AGO had some qualms over the terms. (They also noted that the German army personnel program had dropped the term "psychology" from their vocabulary.) Before the program began, however, the terms were replaced by "personnel consultant" and "personnel technician" – apparently without official action.[23]

Personnel consultants were supposed to have a Ph.D. in psychology and relevant postgraduate experience; personnel technicians were also supposed to have a Ph.D. or one year of graduate training plus relevant experience. It became clear that the available supply of possible candidates was limited, and these high standards were relaxed.[24] The original plan called for a hundred personnel consultants per year. Soon the quota was doubled. In addition to new recruits, already commissioned personnel were identified and trained as personnel consultants. The AGO relied heavily on the Office of Psychological Personnel to help them tag candidates for the program as they entered the military.

In order to retain personnel consultants for longer than the general one-year military service requirement, and to "keep them out of the hair" of commanding officers until they received thorough military indoctrination, selectees underwent an extended period of training. First they were sent to basic training, and then to work in reception and replacement training centers for eighteen weeks. The psychological course came next, followed by Officer Candidate School. Commissioning, because it required an additional year of service, insured that they would stay in the army.[25]

22 Ibid., pp. 321–324.
23 The title of personnel consultant first appeared in a memo on 24 July 1941; ibid., pp. 325–326.
24 In 1942 qualifications for personnel technicians were lowered to a B.A. and preferably two years of graduate work in psychology.
25 Davenport and Kampschroer, *Personnel Utilization,* pp. 307–318.

In June 1942 personnel consultants were officially included in the army tables of organization. According to regulations, their duties were to:

a. Consult with and advise commanding officers on psychological problems.
b. Personally supervise and inspect all phases of psychological work with the organization to which assigned.
c. Recommend assignment of personnel when special psychological problems were involved.
d. Assist the classification officer in the conduct of schools on classification, giving instruction particularly on the administration, scoring and interpretation of tests and other subjects of a psychological nature.
e. This officer may also be a classification and assignment officer.[26]

At first, they worked mainly in induction facilities. When the army began accepting illiterates in August 1942, a personnel consultant was sent to each of the nine corps areas to supervise induction and to coordinate the work of other personnel consultants in the corps area. By November, personnel consultants were directed to advise on all psychological work in command facilities.[27]

Because classification was ultimately the responsibility of the unit commanding officer, it was necessary for the personnel consultants to gain their cooperation in order to be effective. The job was fairly straightforward in induction stations and replacement centers, where the personnel consultant could troubleshoot a variety of personnel problems. In combat units, the consultant had to convince his commanding officer that his services were vital. At the end of the war one observer noted: "In general, personnel consultants accomplished a great deal in non combat units such as induction stations, replacement training centers, special training units and disciplinary barracks. In combat units they were usually unsuccessful." Ironically, and somewhat mysteriously, the personnel consultant job was omitted from the revised army tables of organization published at the end of the war in 1945.[28]

Occupational Skills, Mental Ability, and Physical Fitness

As the army personnel system evolved, competition over manpower allocation developed among the major segments of the army. Classification put a premium on occupational skills and experience, which benefited the Army Service Forces (ASF) in filling their technical and administrative personnel requirements. Public advertisements before the war had propagandized about the vocational value of army service, making specialization more attractive. In contrast, the Army Air Forces (AAF) desired men with high mental ability, in part because aircraft crew skills were relatively rare among civilians. Until 1942 the AAF was able to pro-

26 Quoted from Army Regulations, AR 615-28, 28 May 1942, in ibid., p. 330.
27 Ibid., pp. 339–351.   28 Ibid., p. 351.

cure personnel with high AGCT scores through enlistment alone. Through 1943, when it started having to rely on draftees from reception centers, the AAF obtained permission to take only men with AGCT scores over 100. The army's basic fighting arm, the Army Ground Forces (AGF), was dissatisfied with the system from the start. Voluntary enlistments and generous commissioning policies had drawn many of the best individuals into the navy and marines, and most army volunteers chose the air corps. Only 5% chose infantry or armored force.

The basic complaint of the ground forces was that classification procedures could not measure fighting or combat ability, which was the primary criterion of effectiveness for battle units. The problem was exacerbated because the service force's technical units had first call on new inductees, and the air forces were taking a disproportionate share of the most intelligent. As combat losses mounted and manpower supplies dwindled, the AGF strongly attacked personnel policy in 1943, seeking greater emphasis on the criterion of physical capability in the selection of soldiers. Finally, in the spring of 1944, a simple rating system for physical stamina was instituted. Not surprisingly, psychologists had helped set up a system that reflected their own capabilities and biases. Intelligence and aptitude testing was their strong suit, and their predilection for the role of technical expert reinforced the tendency of the Adjutant General's Office to focus on the tangibles of logistics and organization rather than on the nebulous yet essential characteristics of fighting ability.[29]

The AAF Aviation Psychology Program

Psychologists found even greater scope for their work in the army air forces, in the classification of aircrew personnel. Although the air forces was technically a part of the army, in practice it operated as a self-contained service. Commensurate with the increasing military and political clout of air power during the course of the war, the army air forces gained in status and influence. As part of this expanding institution, the AAF Aviation Psychology Program prospered, gaining a large measure of administrative influence.[30]

Part of the success of the Aviation Psychology Program can be traced to its early establishment in a strategic location by psychologist John C. Flanagan. After receiving his Ph.D. from the Harvard Graduate School of Education in 1934, Flanagan took charge of the annual achievement test development program at the

---

29 Davenport and Kampschroer, *Personnel Utilization,* pp. 31–43. Robert R. Palmer, "Manpower for the Army," *Infantry Journal,* 1947, *61*(1):7–12; (2):27–32; (3):39–43; (4):40–43; (5):39–45; (6):38–45.

30 For a study of ground-crew personnel work, see Victor H. Cohen, *Classification and Assignment of Enlisted Men in the Army Air Arm, 1917–1945* (unpublished manuscript, U.S. Air Force, 1953).

Cooperative Test Service. His position led to wide professional contacts among educators, philanthropists, and psychologists. In 1938 Frederick Osborn, a prominent eugenist, obtained support for Flanagan to conduct a foundation-supported study of family size among air corps pilots. Flanagan's patrons were concerned that pilots – considered the "cream of the crop" genetically – were having small families, and were searching for incentives to encourage them to have more children. Flanagan recruited a young flight surgeon, Captain Loyd E. Griffis, to conduct necessary interviews with air corps personnel. Griffis eventually became Chief of the Research Section, Medical Corps, Office of the Chief of the Air Corps in 1941 and obtained authorization to recruit a "practical psychologist" to develop a psychological research agency. He approached Flanagan, who was restless and ready for a new job. Flanagan received a commission as a major in the Reserves and reported for active duty in July 1941. The Aviation Psychology Program was soon launched, as he recruited thirty-six psychologists to his staff.[31]

The Aviation Cadet Qualifying Examination

The Aviation Psychology Program was administered from the Office of the Air Surgeon at AAF Headquarters in Washington, D.C. The first major task of the program was to develop a basic classification test for aircrew candidates. Flanagan and two colleagues developed the first version of the Aviation Cadet Qualifying Examination, constructing items based on practical aspects of flying. They coordinated their work with the National Research Council's Committee on Selection and Training of Aircraft Pilots. Shortly after Pearl Harbor, the group called a conference with outside consultants to edit and revise the test, and it was put into official use in January 1942.[32]

By early 1942 five field centers were operating. Psychological research units were established at training centers in Nashville, Tennessee, San Antonio, Texas, and Santa Ana, California. The School of Aviation Medicine at Randolph Field, Texas, contained a department of psychology, and Training Command Headquarters at Fort Worth, Texas, had a psychological section in the Office of the Surgeon. The psychological research units were well staffed. Each contained

31 "Distinguished Professional Contribution Award for 1976: John C. Flanagan," *American Psychologist,* 1977, *32:*72–79. The air corps pilot research helped convince its sponsor (the Pioneer Fund) to offer educational scholarships to families having new children; John C. Flanagan, "A Study of Factors Determining Family Size in a Selected Professional Group," *Genetic Psychology Monographs,* 1942, *25:*3–99.

32 Staff, Psychological Branch, Office of the Air Surgeon, Headquarters Army Air Forces, "The Aviation Psychology Program of the Army Air Forces," *Psychological Bulletin,* 1943, *40:*759–769; Staff, Psychological Branch, Office of the Air Surgeon, Headquarters Army Air Forces, "The Aviation Cadet Qualifying Examination of the Army Air Forces," *Psychological Bulletin,* 1944, *41:*385–394.

around twenty officers and a hundred enlisted men; most of the former held advanced degrees in psychology, as did a substantial number of the latter. In addition to administering the testing and classification program, each field center had a special area of research. Psychological Research Unit No. 1 was assigned research on tests of emotion, personality, and temperament; Unit No. 2 on psychomotor tests; and Unit No. 3 on tests of intellectual functions and educational achievement.[33]

In addition to the development of pencil-and-paper tests, the Aviation Psychology Program conducted research on a variety of apparatus tests. Work on the psychomotor test was centered at the department of psychology in the School of Aviation and Medicine and coordinated with Psychological Research Unit No. 2. Operating under the common assumption that perceptual-motor skills are important in flying, the research program adapted existing equipment to the needs of a mass-testing program. For instance, the Automatic Serial Action Apparatus devised by Mashburn in 1934 was refined to produce the SAM Complex Coordination Test, which received widespread use in testing eye–hand coordination.[34]

A related concern with visual cues involved in flying led to the creation of a small Psychological Test Film Unit in 1942. This unit, eventually located in Santa Ana, California, explored the use of motion pictures in testing research and development. Under the direction of James J. Gibson, the unit devised various selection tests and training procedures utilizing motion pictures, coordinating the production of films with the air forces cinema crew in southern California. The small group's practical activities stimulated Gibson's theoretical imagination and contributed to the development of his radical theory of perception after the war.[35]

---

33 For details on the activities and personnel of the Psychological Research Units, see: Staff, Psychological Research Unit No. 1, Army Air Forces, "History, Organization, and Procedures, Psychological Research Units No. 1, Army Air Forces," *Psychological Bulletin,* 1944, *41*:103–114; Staff, Psychological Research Unit No. 2 and Department of Psychology, School of Aviation Medicine, Army Air Forces, "Research Program on Psychomotor Tests in the Army Air Forces," *Psychological Bulletin,* 1944, *41*:307–321; Staff, Psychological Research Unit No. 3, Army Air Forces, "Organization and Research Activities, Psychological Research Unit No. 3, Army Air Forces," *Psychological Bulletin,* 1944, *41*:237–245.

34 Staff, Psychological Research Unit No. 2 and Department of Psychology, Research Section, School of Aviation Medicine, "Research Program on Psychomotor Tests in the Army Air Forces," *Psychological Bulletin,* 1944, *41*:307–321.

35 Staff, Psychological Test Film Unit, Army Air Forces, "History, Organization, and Research Activities, Psychological Test Film Unit, Army Air Forces," *Psychological Bulletin,* 1944, *41*:457–468; James J. Gibson, ed., *Motion Picture Testing Research* (Washington, D.C.: GPO, 1947). On the development of Gibson's theory, see Edward S. Reed, "James J. Gibson's Revolution in Perceptual Psychology: A Case Study of the Transformation of Scientific Ideas," *Studies in History and Philosophy of Science,* 1986, *17*:65–98.

Reorganization

By late 1943 the increasing flood of new recruits prompted a reorganization of the program. Medical and psychological examining units were set up in seven AAF basic training centers to handle testing of candidates without college degrees. The psychological research units continued to examine postcollege recruits. The Training Command Psychological Section gained increased administrative responsibility and utilized new IBM tabulating equipment to keep track of records and to compute statistics.[36]

Beginning in 1944, emphasis in the research program was shifted from classification to aircrew proficiency, training, and combat operations, and new psychological research projects were designated to study the roles of pilot, navigator, bombardier, radar observer, combat crew, and flight engineer. Psychologists also played an important role in the AAF Personnel Distribution Command, which was responsible for evaluating returning combat personnel. Staff members at the six redistribution stations administered instructor selection tests, validated initial classification scores against combat performance, and studied the problems of leadership. Convalescent hospitals were also under the direction of the Personnel Distribution Command, and psychologists were employed in the eleven facilities for testing, counseling, and other clinical activities. Toward the end of the war a psychological branch was activated at the Aero-Medical Laboratory to study aspects of the man–machine system in relation to aviation equipment.[37]

The Aviation Psychology Program, which had employed hundreds of psychologists, demonstrated the utility of a comprehensive research-based personnel program to military officials. The program was based on the theory and methods of factor analysis, particularly those of L. L. Thurstone and T. L. Kelley, whose disciples led the program. Working in this paradigm, researchers used statistical techniques to isolate specific abilities and aptitudes. They identified over a hundred such traits by the end of the war, which gave support to the multiple-factor theories of intelligence. The strong elective affinity between these methods and the needs of the military help account for the success of the Aviation Psychology Program.[38]

36 Staff, Psychological Section, Office of the Surgeon, Headquarters, AAF Training Command, "Psychological Activities in the Training Command, Army Air Forces," *Psychological Bulletin,* 1945, *42*:37–54.

37 Staff, Psychological Branch Office of the Air Surgeon, HQAAF, "Present Organization, Policies, and Research Activities of the AAF Aviation Psychology Program," *Psychological Bulletin,* 1945, *42*:541–552.

38 The program was probably the most thoroughly documented psychology project during the war. The series of nineteen "Aviation Psychology Program Research Reports" were monographs dealing with nearly every aspect of its work. See John C. Flanagan, ed., *The Aviation Psychology Program in the Army Air Forces,* AAF Aviation Psychology Program Research Reports, No. 1 (Washington, D.C.: GPO, 1948); idem, "Research Re-

Personnel Assessment in the OSS

The Office of Strategic Services (OSS), a precursor of the Central Intelligence Agency (CIA), was the setting for an unusual personnel selection program in the Second World War. Organized under the direction of William "Wild Bill" Donovan, the OSS was in charge of foreign intelligence and propaganda activities.[39] Donovan, a military hero in the Great War and a personal friend of President Roosevelt, was a charismatic leader and made the OSS a haven for Eastern intellectuals and foreign émigrés.

A small number of psychologists were employed by the OSS in research activities under Richard Tryon, from the University of California. In October 1943 Tryon, deputy chief of the Planning Staff, was present at an executive meeting when someone suggested creating a personnel assessment for the selection of secret agents patterned after the British War Office Selection Board. The idea was attractive to members of the Schools and Training Branch, which had had problems in their recruitment and selection procedures.[40] A planning group formed, including Tryon and other psychologists, and Donovan authorized the establishment of an assessment unit in late 1943.[41]

The original staff was composed of six psychologists: three OSS members and three newcomers, including Henry A. Murray of Harvard, who emerged as the program's driving force.[42] Both Murray and James A. Hamilton, the unit's administrator, held medical degrees as well as Ph.D.s, setting a pattern of joint psychological-

ports of the AAF Aviation Psychology Program," *American Psychologist,* 1947, *2*:374–375.

39 The OSS was the successor to the Office of the Coordinator of Information, established by President Roosevelt in July 1941.

40 Donald W. MacKinnon, who later helped direct the program, noted that "nobody knew who would make a good spy or an effective guerilla fighter. Consequently, large numbers of misfits were recruited from the very beginning, and this might have continued had it not been for several disastrous operations such as one in Italy for which, on the assumption that it takes dirty men to do dirty works, some OSS men had been recruited directly from the ranks of Murder, Inc., and the Philadelphia Purple Gang." D. W. MacKinnon, "How Assessment Centers Got Started in the United States: The OSS Assessment Program," in *Development Dimensions Monograph I* (Pittsburgh, Pa.: Development Dimensions, Inc., 1974), pp. 1–26, on p. 2.

41 OSS Assessment Staff, *Assessment of Men: Selection of Personnel for the Office of Strategic Services* (New York: Rinehart, 1948), pp. 4–5.

42 Murray's energy and style were legendary. A magazine article on the OSS program described him as "a unique combination of thinking machine, dreamer, two fisted drinker, and scientist." "A Good Man Is Hard to Find," *Fortune,* 1946 (March), *33*(3), p. 93. See also Forrest G. Robinson, *Love's Story Told: A Life of Henry A. Murray* (Cambridge, Mass.: Harvard University Press, 1992).

psychiatric cooperation. The staff grew rapidly, and as in other wartime projects there was considerable turnover. By the end of the war, fifty-seven individuals, including five women, had served on the assessment staff. Over half possessed Ph.D.s in psychology, and twelve were M.D.s (including five dual M.D./Ph.D. holders).[43]

The main assessment center, "Station S," was located outside Washington on a secluded country estate. It conducted a three-day assessment program. In order to handle increasing numbers of recruits, another unit was set up – Station W – in Washington in late winter 1944 that ran a one-day program. Another unit, Station WS, was located on a West Coast beach front. Foreign units were also established to screen native agents in China and other Far Eastern locations and to assess personnel that were being transferred.[44]

The OSS assessment program took a unique approach in evaluating personnel for wartime assignments. In contrast to the military classification programs, which relied on tests measuring discrete aptitudes and skills, the OSS project sought "to see the person whole and to see him real."[45] They emphasized the entire personality of the candidate rather than the possession of discrete skills.[46] This "organismic" orientation jibed well with the nature of OSS job assignments. Agents worked in unusual situations and had to be highly resourceful and self-reliant. And, as nearly all recruits already ranked high in intelligence, traditional testing techniques were not particularly useful in distinguishing suitable personnel. Also contributing to this eclectic approach was the diversity of the assessment staff itself, which included "clinical psychologists of various persuasions, animal psychologists, social psychologists, sociologists and cultural anthropologists, psychiatrists who had practiced psychoanalysis according to the theories of Freud, of Horney, and of Sullivan, as well as psychiatrists who were unacquainted with, or opposed to psychoanalysis."[47] Traditional mental testers and personnel psychologists were conspicuously absent.

Cooperation among this diverse group of specialists was ensured by their enlistment in a common mission and their shared belief in the power of expert observation. Although some attempts were made to formulate a theoretical rationale

43 See OSS Assessment Staff, *Assessment of Men,* pp. v–vii, for a list of staff members.
44 Ibid., p. 5.   45 Ibid., p. 221.
46 Murray was a leading figure in the emerging field of personality research. He built the Harvard Psychological Clinic into a major training and research center. In the 1930s he developed the widely used Thematic Apperception Tests and published an influential text: Henry A. Murray, ed., *Explorations in Personality* (New York: Oxford University Press, 1938); Henry A. Murray et al., *Thematic Apperception Test* (Cambridge, Mass.: Harvard University Press, 1943). See also Rodney G. Triplet, "Henry A. Murray and the Harvard Psychological Clinic, 1926–1938: A Struggle to Expand the Disciplinary Boundaries of Academic Psychology" (Ph.D. diss., University of New Hampshire, 1983).
47 OSS Assessment Staff, *Assessment of Men,* p. 26.

for the assessment program, the task was abandoned as impossible given the diversity of opinion and the lack of time to reach consensus. Instead the staff concentrated on developing principles and procedures of assessment. Ideally, the system was based on eight major guiding principles or steps, beginning with the determination of job requirements and qualifications, through the use of a battery of situational tests, to the writing of a personality sketch and its discussion at a staff conference. As the staff characterized it:

> The scheme employed by us might be called the multiform organismic system of assessment: "multiform" because it consists of a rather large number of procedures based on different principles, and "organismic" (or "Gestalt," or "holistic") because it utilized the data obtained through these procedures for attempting to arrive at a picture of personality as a whole; i.e., at the organization of the essential dynamic features of the individual. The knowledge of this organization serves as a basis both for understanding and for predicting the subject's specific behavior.[48]

Test Procedures

The assessment program at Station S took three days. Because of the security requirements, the candidates were given student names and were required to create cover stories to mask their identities, the convincing maintenance of which became a part of their assessment. Upon their arrival in the evening, the candidates would fill out a lengthy personal history form, take a variety of objective and projective paper-and-pencil tests, and complete questionnaires on health conditions and work preferences.

The bulk of assessment consisted of an array of situational tests. The recruits were observed performing a variety of group as well as individual tasks. Some were outdoor situations, such as the wall test where a small group of men would have to cooperatively scale a series of wooden barriers. The construction test, in which a candidate was instructed to build a simple wooden structure with the "aid" of two obstreperous confederates, was designed to measure tolerance for frustration. Other tests were designed to measure verbal resourcefulness in group discussion and skills in preparing written and verbal propaganda. A simulated enemy interrogation the "stress interview" – proved to be highly revealing of emotional stability.

One test situation, called improvisations, was based loosely on Moreno's psychodrama techniques and evolved into an important method of predicting reactions to varied situations. After a few scenes were enacted, the participants were encouraged to drink alcoholic beverages, which added another variable to be studied.

During the entire testing program, OSS psychologists and psychiatrists were

48 Ibid., p. 28. Principles of assessment are discussed in detail on pp. 26–57.

constantly evaluating the candidates. Their close involvement made them "more than mere testers and scorers." Rather, they were "part of the situation itself, components of the differentiated social world with which each candidate had to deal." In the final analysis, assessment relied on their clinical judgments. In formulating their evaluations, the OSS staff conceptualized their work in terms of clinical medicine. Three major steps were involved: "(1) obtaining data (observation and scoring tests); (2) forming a unified conception of the personality (diagnosis); (3) estimating the probable level of future performances (prognosis)."[49]

Test scores and ratings were compiled onto a central rating chart and formed the basis for the personality sketch. The personality sketch was revised throughout the assessment period and became the key item in candidate evaluation. Final disposition of each case was accomplished in a staff conference at the end of the three-day period in which all staff members would share their findings and opinions on the suitability of candidates for their proposed assignments.[50]

Program Rationale and Results

The OSS assessors viewed their program as an experiment in social psychology. By disguising the students' identities, they believed they were creating a temporary "classless" society where status was achieved rather than ascribed. Although the staff had a clearly defined evaluative role, they interacted closely and almost continuously with the candidates. The notion that social class was being minimized was no doubt reinforced by the fact that staff and students shared many important social characteristics: they were predominantly young, male, white, native-born, and highly educated. The assessors, however, were not blind to their own potential for bias, noting that Station S staff "were no doubt unconsciously disposed to overrate men who enjoyed and entered into the spirit of a program which they had designed and were administering."[51]

Validation problems, inherent in all wartime personnel procedures, plagued the OSS program. The primary criterion of success – job performance – was vague and often impossible to measure. Men were given assignments different from those for which they had been selected; the assignments were often poorly defined; their activities in the field were difficult to observe; and what counted as success varied considerably depending on the evaluator. Writing after the war, the staff admitted, "thus there is no tangible proof that the OSS assessment staffs produced effects which more than balanced the expenditure of time and money."[52] But the program was generally viewed as a great improvement over the previous selection system, and incomplete medical records of OSS personnel suggested that the incidence of psychiatric breakdown among screened individuals was low.

In the postwar account of the program in *The Assessment of Men,* the staff in-

49 Ibid., p. 203.   50 Ibid., pp. 212–221.   51 Ibid., p. 458.   52 Ibid., p. 451.

cluded a section on "The Promise of Assessment" that suggested directions for research and application. Reiterating enduring themes of cultural lag intensified by the advent of the atomic bomb, the report warned apocalyptically: "material science has taken on the character of a cancerous growth, and, if not balanced by the development of a usable social science *operating in the service of humanistic values,* it will surely pass from the state of doing man more good than harm to that of doing him more harm than good, if not of demolishing his most valued institutions."[53] They recommended that a research institute or agency be established to build on the wartime program. In addition to selecting armed forces personnel, business executives, or high-level government officials, the proposed institute would be used to train clinical psychologists and psychiatrists as well as for research.[54] As these recommendations were being written, some of Murray's ideas were being incorporated into the new Harvard Department of Social Relations, which included clinical and social psychology, sociology, and cultural anthropology. The creation of "SocRel," by including the Psychological Clinic among its facilities, gave Murray's program the institutional status and recognition it had been denied for years by the Harvard administration and the psychology department.[55] The OSS program's strong connections to the University of California resulted in the creation of the Institute of Personality Assessment and Research in 1949. Donald MacKinnon, who headed the Station S staff, was attracted to California in 1947 by Tolman and Tryon and became the institute's first director.[56]

53 Ibid., pp. 463–464.   54 Ibid., pp. 462–473.
55 Gordon W. Allport and Edwin G. Boring, "Psychology and Social Relations at Harvard University," *American Psychologist,* 1946, *1:*119–122; Talcott Parsons, *Department and Laboratory of Social Relations, Harvard University: The First Decade, 1946–1956* (Cambridge, Mass.: Harvard University, 1957). On Murray's struggles before the war, see Triplet, *Henry A. Murray and the Harvard Psychological Clinic.*
56 Joan E. Grold, "A History of the University of California Psychology Department at Berkeley" (unpublished paper, University of California, 1961), pp. 1–47, on p. 26. MacKinnon notes that Michigan Bell Telephone Company established the first operational assessment center in industry around 1958; MacKinnon, "How Assessment Centers Got Started," p. 23.

5

Applied Human Relations
The Utility of Social Psychology

Three months after the creation of the Emergency Committee in Psychology in August 1940, the National Research Council sponsored a Conference on Psychological Factors in Morale. Twenty-five psychologists attended the two-day meeting, held in Washington. Gordon Allport, a social psychologist at Harvard who served as APA president in 1939, chaired the sessions. Organizational problems dominated the agenda, particularly the relation of the Emergency Committee to morale work. Discussion revealed a generational split, with younger members of the conference expressing some misgivings about organizing under the wing of the newly formed Emergency Committee. Allport thought the issue of morale merited wide consideration: "is not morale so gigantic a problem that it ought to feature large in the work of the NRC and not be located as a tail to the Emergency Committee's kite?" He argued that it might be best to "assemble a mixed group of social scientists" as a subcommittee.[1]

Allport, slightly senior to the younger group of social psychologists, was well aware of his role as a representative of their interests in the psychological establishment. In addition to his prominent Harvard perch, he had been the first APA president to be identified with the field of social psychology. Before the conference, he had been involved in the Committee for National Morale, a civilian initiative organized by Arthur Upham Pope, a Persian art expert who had been active in similar efforts during the First World War. Pope had contacts among psychologists through his prior service on the army's Committee for the Classification of Personnel. Armed with a 500-page proposal, he petitioned President Roosevelt to establish a federal morale agency.[2]

Many psychologists lent their names to the effort, including Allport, James

---

1 James H. Capshew, "Psychology on the March: American Psychologists and World War II" (Ph.D. diss., University of Pennsylvania, 1986), p. 166.
2 Ibid., p. 163.

Angell, Walter Bingham, Hadley Cantril, Leonard Carmichael, Leonard Doob, Kurt Lewin, Gardner Murphy, Henry Murray, Robert Yerkes, and Goodwin Watson. Plans were drawn up for a "psychological research unit" in the proposed Federal Morale Service that would provide a clearinghouse for morale work, act as a liaison between the government and social scientists, supply the government with confidential information on morale, and perform long-range research projects "vital to the well-being of the people." Among the first projects sponsored by the Committee for National Morale was the preparation of a book on *German Psychological Warfare* based on a comprehensive bibliography of the subject. Allport coordinated the project from his Harvard office. First published in June 1941, the book was sold to raise money for the committee.[3]

After the Conference on Psychological Factors in Morale, Allport requested funds from the Carnegie Corporation to help set up a morale committee under the National Research Council. At the suggestion of Emergency Committee member Walter Hunter, an experimental psychologist from Brown University, he changed the proposal's title from "Morale" to the more scientific-sounding "Social Motivations and Affectiveness." Hunter, more of an insider in national scientific circles than Allport, gave only qualified support to Allport's plans while he pursued a different agenda through the NRC. Because of widespread interest in the morale problem, Allport was asked to withdraw his Carnegie proposal in favor of the joint activity of the NRC, the Social Science Research Council, the American Council of Learned Societies, and the American Council on Education. In June 1941, Hunter was asked to chair a new joint committee on Problems of Morale. Writing to Allport with the news, Hunter exclaimed, "I could have been knocked over with a feather!" After spending weeks investigating the possibilities, Hunter not only declined to head an intercouncil committee but recommended against its formation, arguing that such a committee would be unnecessary because the army already had a morale branch and civilian government agencies were independently organizing their own research in the area. Furthermore, he believed that the NRC could handle any additional requests without a special committee.[4]

Meanwhile Allport's Harvard office had become an unofficial clearinghouse for morale studies as well as a communications center for social psychologists seeking national service. Although he was offered jobs in government agencies, Allport decided to remain at Harvard, aiding the war effort without becoming a regular federal employee.[5]

3 Ibid., pp. 163–164. Ladislas Farago, ed., *German Psychological Warfare: Survey and Bibliography* (New York: Committee for National Morale, June 1941; 2d ed. September 1941; reprinted by G. P. Putnam's Sons, New York, 1942).
4 Capshew, "Psychology on the March," pp. 166–167.
5 Ibid., pp. 167–168.

Subcommittee on Defense Seminars

As the representative of the Society for the Psychological Study of Social Issues, Allport sat on the Emergency Committee in Psychology. After a report by psychologists working in the Office of the Coordinator of Information (soon renamed the Office of Strategic Services), the Emergency Committee appointed a Subcommittee on Defense Seminars in November 1941, with Allport as chair. This action provided formal recognition of the role Allport's office was playing as a national clearinghouse. The idea was that his office would receive problems in social psychology from government agencies and allocate them among seminars organized in various universities. The seminars were modeled upon the ones already being conducted at Harvard by Allport and his colleagues. Following a familiar academic format, faculty and graduate students surveyed the published literature, devised empirical studies, and wrote summary reports. The idea caught on quickly, and by January 1942 over twenty seminars were being held around the country.[6]

The subcommittee had tapped a reservoir of public concern and was soon inundated with requests and suggestions, which poured in even faster after Pearl Harbor. Allport and his assistant Gertrude Schmeidler described the response:

> Too many psychologists in too many institutions wanted to help. Isolated individuals, undergraduate classes, local groups of various sorts became correspondents of the Subcommittee. Psychologists wanted to know how to establish Rumor Clinics, how to clear the findings of their local investigations of morale, where to place patriotic radio programs they had written – a thousand and one things – above all what they could do to help in the war effort. Meanwhile, government requests began to come in from a great variety of agencies. . . .

As psychologists continued to leave universities for government employment, the formal seminars gave way to efforts by individuals or informal groups. In December the NRC authorized Allport to approach the Social Science Research Council for additional funding for the subcommittee.[7]

A Clearinghouse for Social Psychology?

The subcommittee's increasing prominence as a general clearinghouse for social psychology reactivated controversy over the Emergency Committee's relation to social psychology. Some Emergency Committee members regarded the creation of the subcommittee "as an encroachment or trespass upon the field of the Social Science Research Council," according to chair Karl Dallenbach, who was speaking for the conservative experimentalists. Given psychology's affiliation with both

6 Ibid., p. 168.   7 Ibid., pp. 168–169.

the NRC and the SSRC (the American Psychology Association had long-standing ties to both councils) and the fact that the SSRC had maintained its traditional low-key advising role even during mobilization, the objections reflected ideological differences among committee members rather than a potential jurisdictional dispute between the councils. The real problem was antipathy in the NRC toward involvement in "social" (i.e., political) issues. Social psychology, although it had many strong supporters, could boast only a few prominent practitioners, and the field had not yet attained respectability as a psychological specialty. In mainstream groups like the Emergency Committee, social psychologists were under continuing pressure to justify their work as part of scientific psychology.[8]

At the May 1942 Emergency Committee meeting Allport requested funding for secretarial services and office expenses, and suggested that the subcommittee be redesignated as a standing Committee of Social Psychology. In seeking to bypass the biased Emergency Committee and secure an organizational base for social psychology, Allport had made a tactical blunder by drawing NRC officials into the debate. While a majority of Emergency Committee members could be persuaded to continue support for the subcommittee's work, NRC officials were less sympathetic. They immediately rejected the proposal for a new committee, scrupulously avoiding any possible encroachment on SSRC territory. The SSRC also rejected Allport's idea. Exhibiting the council's typical reluctance to take the initiative, SSRC officials claimed that if social psychology got a clearinghouse then all the social sciences would want one, and that "if the government wanted a clearinghouse then it would provide it."[9]

After several further unsuccessful attempts to gain funding for the subcommittee, Allport recommended that it be discharged at the September Emergency Committee meeting and tendered his own resignation as SPSSI representative as well. After resigning from the Emergency Committee in 1942, Allport channeled more energy into SPSSI's wartime activities, such as its yearbook *Civilian Morale*. Soon afterward he was selected as chair of its new Committee on War Service and Research. Allport continued his volunteer services in organizing social psychology and became the center of a large network of civilian researchers on morale. With the aid of students and colleagues, he was able to generate, collect, and disseminate large amounts of material on various aspects of morale, including measurement, public opinion methods, minorities, rumor, morale building, media of communication, propaganda, comparative national psychology, industrial morale, demoralization, children, and postwar planning.[10]

8 Ibid., pp. 169–170.   9 Ibid., p. 170.

10 Ibid., pp. 171–172. See Gordon W. Allport and Helene R. Veltfort, "Social Psychology and the Civilian War Effort," *Journal of Social Psychology,* 1943, *18:*165–233; Gordon W. Allport and Gertrude R. Schmeidler, "Morale Research and Its Clearing," *Psychological Bulletin,* 1943, *40:*65–68; Gertrude R. Schmeidler & Gordon W. Allport, "Social

Surveying Attitudes and Opinions

Although Gordon Allport's efforts to establish an overall policy for the mobilization of social psychology failed, social psychologists, as individuals and in groups,
were successful in finding a number of outlets for their skills in the war effort, particularly in government-sponsored survey research. Federal officials, both military
and civilian, found knowledge of American and foreign attitudes and opinions to
be useful in making decisions and supported an array of survey research groups.[11]

The Division of Program Surveys

The U.S. Department of Agriculture's Division of Program Surveys was established in 1939 with the object of gaining knowledge about farmers' attitudes that
would be useful in administering farm relief. It was headed by social psychologist
Rensis Likert, who had received his doctorate at Columbia in 1932 under Gardner
Murphy and then applied his innovative survey methods in a variety of commercial and academic settings before going to the U.S.D.A. With its staff of field interviewers and under Likert's entrepreneurial leadership, the division was strategically positioned to provide the government with information on civilian attitudes
and opinions. The group demonstrated its resourcefulness when it conducted interviews with Los Angeles residents on the day following Pearl Harbor and delivered a report on civilian reactions to the attack within a week.[12]

    The division expanded quickly, growing from an original staff of 12 to more
than 200 employees by 1942, including nearly 75 field interviewers. Funding
tripled in a single year, from $99,000 in 1941 to $332,000 in 1942. Much of the
money came from contracts with the Office of Facts and Figures, a government
agency charged with the task of providing public information regarding the defense effort. After Pearl Harbor, government information programs were consoli-
dated under the Office of War Information, which continued to contract research
from the Division of Program Surveys.

    In contrast to pollsters, who asked simple questions of an extensive sample of

Psychology and the Civilian War Effort: May 1943–May 1944," *Journal of Social Psychology,* 1944, *20:*145–180; Gordon W. Allport and Leo Postman, *The Psychology of
Rumor* (New York: Henry Holt, 1947).
11 For general background, see Wroe Anderson, "Trends in Public Opinion Research," in
    A. B. Blankenship, ed., *How to Conduct Consumer and Opinion Research* (New York:
    Harper, 1946), pp. 289–309. See also Jean M. Converse, *Survey Research in the United
    States: Roots and Emergence, 1890–1960* (Berkeley: University of California Press,
    1987).
12 This section relies on Luis R. P. Blanco, "Rensis Likert and the Transformation of Sam-
    ple Survey Methodology" (B.A. honors thesis, Department of History and Sociology of
    Science, University of Pennsylvania, 1986), p. 26.

people, Likert's group based its work on intensive interviews with selected individuals. They wanted to determine the underlying attitudes of their subjects rather than general shifts in expressed opinion. As the Office of War Information developed into a government news service, however, polling techniques came to be favored over the intensive interview method. As a result, the division's contracts were terminated by early 1943 and the group had to look elsewhere for funds. It succeeded in obtaining contracts from the Federal Reserve Board and other agencies. Division members also moved to other wartime projects.[13]

The War Production Board

The economic disruptions of wartime also provided some new employment opportunities. A small group of psychologists were hired by the War Production Board in connection with its national consumer surveys. As the central agency charged with accelerating manufacturing and industrial production, it had authority to determine priorities and allotments of materials pertaining to war production. By the spring of 1943, with the shift to a wartime economy, consumer shortages began to be felt. Complaints and protests about shortages poured in from citizens, Congress, and manufacturers, and the board set up a Civilian Relations Division (later renamed Civilian Surveys Division) in the summer of 1943 to determine the nature and extent of consumer requirements. Interviewers were sent into the field, and their reports indicated that serious problems were developing. Accurate estimates of shortages and requirements were needed, and a nationwide survey was planned. Survey researchers, including George Gallup and Paul Lazersfeld, were brought in as consultants, and four psychologists recruited from Likert's group – Dwight Chapman, Ernest Hilgard, Helen Peak, and Ruth Tolman – comprised most of the senior staff of the survey program.[14]

The survey schedule began with an open-ended question about shortages that had most bothered the respondent. Then a series of questions concerning a list of 115 items of goods and services was asked in order to ascertain difficulties in obtaining the items. Rather than devise their own national sample, the surveyors used a master sample of 5,000 households developed by the Bureau of the Census for

---

13 Ibid. See also Harold F. Gosnell and Moyca C. David, "Public Opinion Research in Government," *American Political Science Review,* 1949, *43:*564–572. Psychologists were also employed in other branches of the Office of War Information; for example, see Leonard Doob, "The Utilization of Social Scientists in the Overseas Branch of the Office of War Information," *American Political Science Review,* 1947, *41:*649–667.

14 Ernest R. Hilgard, "Psychologists in the War Production Board," 4 December 1945, 6 pp; APA/G-8/Yerkes Committee, Miscellaneous Reports; Charles E. Noyes and Ernest R. Hilgard, "Surveys of Consumer Requirements," in A. B. Blankenship, ed., *How to Conduct Consumer and Opinion Research* (New York: Harper, 1946), pp. 259–273.

its monthly labor report. Interviews were conducted by the census bureau's field staff in November 1943 and the results tabulated by machine. By December a report, the "First Survey of Consumer Requirements," was available to policy-makers.[15]

Other surveys of civilian consumer behavior were conducted, and it became clear that such new quantitative data provided a better basis for the prediction of demand than prewar production and consumption figures. In September 1944 the chair of the War Production Board announced that civilian production programs would cease after the anticipated victory in Europe and the survey division began to wind down. One last national survey was performed, in April 1945, which canvassed previous categories as well as consumer intentions for postwar purchases.[16]

Attitude and Opinion Research in the Army

Under the sponsorship of Frederick Osborn, head of the army's Division of Education and Information (formerly the Morale Services Division), sociologist Samuel Stouffer assembled an impressive group of social scientists to measure the morale and opinions of U.S. soldiers as an aid in formulating policy. The Research Branch under Stouffer was divided into two sections. The Survey Section, staffed mainly by sociologists, performed large-scale cross-sectional surveys of the army population. The smaller Experimental Section, composed primarily of psychologists, conducted experiments on techniques of mass communication, including the use of films and other media.[17]

The mission of the Experimental Section was "to make experimental evaluation of various programs of the Information and Education Division." Much of its work dealt with the series of "Why We Fight" indoctrination films made by Frank Capra, the famous Hollywood director. Other media were examined, such as newspaper and book readership and radio listening habits, and research was performed for other parts of the War Department.[18]

Yale psychologist Carl Hovland was head of the section. Born in 1912, he was

15 Ibid. The survey generated a technical study of differences between households easy and hard to find at home; Ernest R. Hilgard and Stanley L. Payne, "Those Not at Home: Riddle for Pollsters," *Public Opinion Quarterly,* 1944, 8:254–261.

16 Noyes and Hilgard, "Surveys of Consumer Requirements," pp. 264–270.

17 For an analysis of the Survey Section's wartime work and its postwar consequences see Peter Buck, "Adjusting to Military Life: The Social Sciences Go to War, 1941–1950," in M. R. Smith, ed., *Military Enterprise and Technological Change: Perspectives on the American Experience* (Cambridge, Mass.: MIT Press, 1985), pp. 205–252.

18 Carl I. Hovland, Arthur A. Lumsdaine, and Fred D. Sheffield, *Experiments on Mass Communication* (Princeton, N.J.: Princeton University Press, 1949), pp. v–vii, quote on p. v.

among the oldest members of the eight-person staff, whose average age in 1942 was less than twenty-eight. Only Hovland and one other staff member held Ph.D.s; the rest were graduate students in psychology. The group did share some significant ties: nearly all had studied at Yale, Stanford, or the University of Washington beginning in the mid-1930s.[19] The section's advisors included Quinn McNemar, a psychologist engaged in study of attitude and opinion research methodology for the Social Science Research Council, and sociologists John Dollard, Paul Lazarsfeld, and Robert Merton.[20]

The section's findings, which were presented in the third volume of *Studies in Social Psychology in World War II,* were based mainly on studies involving motion pictures. The research utilized experimental as well as evaluative techniques and differed from most commercial film research through its emphasis on the measurement of behavioral and cognitive changes in the audience rather than simple approval ratings. By analyzing the effects of films on soldiers, the psychologists hoped to discover principles that would apply to mass communications in general. Three main classes of variables were studied: population variables (e.g., age, intelligence, etc.), film content, and external variables (e.g., presentation of supplementary material). Film effects were measured by audience evaluations and reactions, and by testing what the audience had learned. Among the section's first tasks was to evaluate the series of "Why We Fight" documentary films, which were designed to replace orientation lectures at training camp. As instructional and training films became more widely used in the army, the section was called upon to judge their usefulness. Finally, audience reactions to general-interest films were also studied.[21]

The Strategic Bombing Survey

In November 1944 Likert was appointed head of the Morale Division of the recently established U.S. Strategic Bombing Survey. With President Roosevelt's approval, the secretary of war had authorized a wide-ranging inquiry to determine the effects of Allied bombing campaigns on Germany. The major thrust of the survey was economic and industrial analysis, although studies of political, medical, and psychological effects were also included. The survey, directed by civilians,

19 Staff members were: F. J. Anderson, J. L. Finan, C. I. Hovland, I. L. Janis, A. A. Lumsdaine, N. Maccoby, F. D. Sheffield, and M. B. Smith. Several other persons worked for the Section for shorter periods: J. M. Butler, D. A. Grant, D. Horton, E. H. Jacobson, A. Marblestone, A. H. Schmid, and A. Turetsky.
20 Ibid., p. vii.
21 Ibid., pp. 3–16. For a retrospective review of some lines of research initiated by the section, see Arthur A. Lumsdaine, "Mass Communications Experiments in Wartime and Thereafter," *Social Psychology Quarterly,* 1984, *47:*198–206.

was supported by Air Force leaders as a way of defining a distinctive mission – "strategic bombing" – and as a basis for postwar planning.[22]

Likert's high visibility landed him the job, and he recruited his staff from the Division of Program Surveys and from his wide contacts among social scientists.[23] He attracted a number of psychologists to serve, including Otto Klineberg, David Krech, Robert MacLeod, Theodore Newcomb, and Helen Peak. Among other notable staff members were political scientist Gabriel Almond and British poet W. H. Auden. By March 1945, Likert had forty military and civilian personnel in his unit, and they began to prepare for fieldwork by examining captured German letters and other documents.

As soon as Germany surrendered in April, the Morale Division went to work, setting up its first headquarters in Darmstadt and then in Bad Nauheim. The division employed over 130 people at its peak, and sent a total of 14 teams into the field. Based on the results of nearly 4,000 interviews, the group found little evidence that German civilian morale had been adversely affected by Allied bombing, at least not until defeat was imminent.[24]

As the European fieldwork was ending in the summer of 1945, plans were made to perform a similar survey for Japan. Likert was unable to make the trip, so a younger colleague, Burton Fisher, was made head of the Morale Division. Although there was some overlap with the personnel of the German survey, many new staff members were added, most notably a number of anthropologists (e.g., Conrad Arensberg, Alexander Leighton). With over 200 employees, the Morale Division was nearly the largest unit of the Strategic Bombing Survey in Japan. The group conducted over 3,000 interviews during the last three months of 1945. The massive amount of data was coded at Swarthmore College under the direction of David Krech and with the aid of numerous students.[25]

Defining morale as a "short-hand term for a complex of factors which indicate

22 David MacIsaac, *Strategic Bombing in World War Two: The Story of the United States Strategic Bombing Survey* (New York: Garland, 1976).
23 Evidently Likert was recommended by Leonard Carmichael when he was asked for suggestions by James Bryant Conant, Harvard president. MacIsaac, *Strategic Bombing,* p. 188n9.
24 James Beveridge, *History of the United States Strategic Bombing Survey,* 4 vols., Washington, D.C., July 1946; vol. 1, pp. 360–377; USSBS/ACMH/4. U.S. Strategic Bombing Survey, Morale Division, *The Effects of Strategic Bombing on German Morale,* 2 vols. (Washington, D.C.: U.S. Strategic Bombing Survey, 1947).
25 Krech was brought to Swarthmore by his Morale Division colleague, Robert MacLeod, psychology department chair. Beveridge, *History of the United States Strategic Bombing Survey,* vol. 1, pp. 189–202. See also a participant's account: George H. H. Huey, "Some Principles of Field Administration in Large-Scale Surveys," *Public Opinion Quarterly,* 1947, *11:*254–263.

the willingness and capacity of the Japanese to follow their leaders and to work and sacrifice to win the war," the division found air attacks to be a major factor in the decline of morale among Japanese civilians after late 1944. This was hardly surprising, given the extent of American bombing: an average of 43% of each of Japan's sixty-six largest cities had been destroyed, and over one-third of the civilian population had personally experienced bombing raids. The results also indicated that the atomic bombings had not caused Japan's surrender, but only hastened it.[26]

Professionalizing Social Psychology

The war helped to strengthen the already dense web of connections among social psychologists. Prewar educational and professional ties were augmented by wartime deployment patterns that favored multiple job positions and interdisciplinary cooperation. Organizations like the Society for the Psychological Study of Social Issues were mobilized to facilitate communication and encourage action. Informal groups sprang up spontaneously. For instance, a discussion group coalesced in Washington in early 1942, as social psychologists streamed into the city to take up their governmental duties. Social psychologist Steuart Henderson Britt, head of the Office of Psychological Personnel, helped organize the group and arranged for its biweekly meetings. Focusing on the role of social psychology in the war effort, the group heard presentations by Leonard Doob, Rensis Likert, Otto Klineberg, Theodore Newcomb, Goodwin Watson, and others concerning their activities. Broader professional issues were raised by other psychologists, such as Robert Yerkes, who spoke on "The Professionalization of Psychology," and Ernest Hilgard, who commented on "Post-War Planning for Psychology." The meetings, which continued into the second year of the war, attracted around two dozen psychologists and provided an important avenue of communication and fellowship.[27]

The war had a profound impact on the Society for the Psychological Study of Social Issues. It provided a perfect opportunity for SPSSI members to put their ideals into practice, and with the failure of other attempts to organize social psychology the group became a major center of wartime activity. During the war SPSSI continued its practice of publicly addressing current social issues through various publications. In addition to the topical yearbook on *Civilian Morale* (1942), the society disseminated a list of ten suggestions offering "propaganda protection" for American citizens. Including such dictums as "Don't think of governments as

---

26 U.S. Strategic Bombing Survey, Morale Division, *The Effects of Strategic Bombing on Japanese Morale* (Washington, D.C.: U.S. Strategic Bombing Survey, 1947); MacIsaac, *Strategic Bombing in World War Two,* pp. 115, 138, 207n32.

27 Capshew, "Psychology on the March," pp. 188–189.

having personalities" and "Don't trust emotional phrases," the SPSSI warnings were a commonsensical form of mental hygiene. When a booklet on "The Races of Mankind" written by anthropologist Ruth Benedict and Gene Weltfish of Columbia University was condemned as subversive by Congress and threatened by efforts to block its distribution, the SPSSI Council voted to release a statement in support of the booklet and its message "that there is no real evidence for the belief in the innate superiority of any one race over any other, and that racism is therefore superstition." Gordon Allport, as chair of the society, headed the list of signers, which included a number of members active in war work.[28]

SPSSI members, like American psychologists generally, began thinking about the postwar world almost from the start of U.S. involvement. In the middle of the war Gardner Murphy was appointed head of a committee to produce the next SPSSI yearbook, which was to deal with the problems of securing a lasting peace. In addition to the usual short articles, plans for the book included a manifesto on "Human Nature and the Peace." Allport was in charge of drafting the statement and gathering signatures for it. By the summer of 1944 he was circulating a draft version to his colleagues. Although most agreed with its intentions, a few prominent psychologists would not sign it. Robert Yerkes declined because he felt the statement was based on social "desiderata" rather than on psychological principles. He asserted, "as a comparative psychologist I know that various aggressive tendencies are built into some organisms, including some men!" Walter Hunter also claimed to have sympathy with the statement's aims when he declined to sign, but he was more concerned about its impact on psychology's public image if it was circulated at the fall meetings of the American Psychological Association and the American Association for the Advancement of Science. Hunter's misgivings notwithstanding, the manifesto was circulated at the APA meeting, and psychologists readily signed it. Some two thousand signatures were gathered (over 50% of APA membership), and only twelve psychologists expressed their dissent. The statement was included in the SPSSI yearbook, *Human Nature and Enduring Peace,* published in 1945.[29]

The varied demands of war created opportunities for social psychologists to work together on large-scale cooperative projects, which fostered solidarity among themselves and new relations with other social scientists. For example, the Strategic Bombing Survey brought together a number of leading social psychologists. As David Krech commented later: "Likert's USSBS study seemed to be wholly a SPSSI operation. Among those present in Germany were Dorwin Cartwright, Daniel Katz, Herbert Hyman, Richard Crutchfield – every one an officer (then or later) of SPSSI. Many an evening was spent in Darmstadt and Bad Nauheim plotting and planning the postwar future of social psychology in America and the role SPSSI could play." The fact that few psychologists worked at the same job over

28 Ibid., pp. 189–190.   29 Ibid., pp. 190–191.

the entire course of the war increased the likelihood of new contacts and encouraged identification with the discipline rather than with a particular institutional context.[30]

By providing governmental sanction for the work of SPSSI members, the war helped to move the organization more firmly into the mainstream in psychology. The rhetoric of social utility espoused by SPSSI leaders resonated harmoniously with similar expressions voiced from other quarters of the psychology community. Goodwin Watson, SPSSI's founding chairman, expressed the hopes and aspirations of his colleagues in a fanciful article on "How Social Engineers Came to Be," which pretended to be the presidential address at the fortieth anniversary meeting of the American Social Engineering Association in 1983. Tracing the roots of the profession to the social planning of the Great Depression and the development of survey research, Watson emphasized the importance of the war in "lur[ing] psychologists away from their cages of white rats" and toward the consideration of social behavior. In his hypothetical account, social engineering had become so successful that new subspecialties were threatening to fragment the profession. Watson warned:

> If the various specialized interests go off in their separate ways, we shall lose the unity which has previously been very important in building for our profession the high prestige which it today enjoys. We want to remain united in our faith in scientific methods and in our judgment that human and spiritual values take precedence over the material instruments of life.

In the guise of prophecy, Watson was expressing contemporary sentiments favoring a united scientific profession.[31]

Such views articulated nicely with the ambitions of other segments of the psychology community to create a technoscience. Although each group had specific concerns and goals for its own field, they shared a common faith that psychology could help solve personal and social problems. SPSSI members were deeply involved in the reconstitution of the American Psychological Association during the war, and the society voted to become a constituent group – Division 9 – of the new APA. After the war Dorwin Cartwright commented: "Just as the first World War witnessed the establishment of psychological testing as a major field of psychology, it now appears that the second World War has brought to maturity social psychology."[32]

---

30 Ibid., pp. 191–192; David Krech, "Autobiography," in *History of Psychology in Autobiography,* vol. 6 (New York: Prentice-Hall, 1974), 221–250, quoted on p. 243.
31 Goodwin Watson, "How Social Engineers Came to Be," *Journal of Social Psychology,* 1945, *21:*135–141, quoted on pp. 137, 141.
32 Ibid., p. 193; Dorwin Cartwright, "Social Psychology in the United States during the Second World War," *Human Relations,* 1947, *1:*333–352, quoted on p. 333.

6

From the Margins
Making the Clinical Connection

In 1940, Chauncey McKinley Louttit (1901–1956), a clinical psychologist from Indiana University, sought and received a commission from the U.S. Navy. As the navy's first psychologist, he hoped that war work would provide a way out of his professional frustrations. Louttit had done much to advance the cause of clinical psychology during the 1930s and was eager to broaden his already extensive networks. The author of a standard textbook in the field, published in 1936, he was involved in the founding of the American Association for Applied Psychology in 1937 and was serving as executive secretary of the group.

Louttit, trained as a comparative psychologist under Robert Yerkes in the late 1920s, was ambivalent about making a career in clinical psychology and entered the field reluctantly after graduate school as a temporary employment expedient. His career illustrates the marginal status of clinical psychology before World War II.

Louttit's first paid employment in psychology prefigured his later career. After receiving his bachelor's degree from Hobart College in 1925, he became an assistant to Stanley D. Porteus, director of research at the Vineland Training School for the Feebleminded, in New Jersey. Earlier, under Henry H. Goddard, Vineland had become one of the first sites for the use of psychological tests for the diagnosis and treatment of mental problems.[1] After several months, Porteus recommended Louttit to Robert Yerkes for a graduate assistantship at Yale. The timing was fortunate. Yale had just established its Institute of Psychology with funds from the Rockefeller Foundation, and Yerkes was launching his ambitious research program in comparative psychology. With fellowship money from the National Research Council's Committee on the Problems of Sex, Yerkes was searching for a student to "devote a few years, if not his life, to the study of fundamental problems of sex."

---

1 See Leila Zenderland, *Measuring Minds: Henry H. Goddard and the Development of Mental Testing in America, 1908–1918* (New York: Cambridge University Press, forthcoming).

Louttit won the fellowship after expressing his eagerness to learn psychobiology and saying that he found research more interesting than clinical work.[2]

Yerkes set Louttit to work on the sexual behavior of the guinea pig. His dissertation, completed in 1928, was published as a monograph in the *Journal of Comparative Psychology,* which Yerkes edited. Louttit also demonstrated a flair for library research, and with Yerkes's encouragement published the massive "Bibliography of Bibliographies in Psychology: 1900–1927."[3] After graduating Louttit landed his first job, at the University of Hawaii Psychological Clinic, through Porteus, who had become its director. Cut off from the heady atmosphere of Yale, Louttit soon grew disappointed with the position, which consisted largely of routine testing of schoolchildren. He decided, however, to stay for two years out of loyalty to Porteus, and then go someplace "more scientific." As he confided to a graduate school friend: "I realize the experience gained here will not be valueless, especially on the clinical side. I am not interested in clinical work, but the knowledge may come in handy."[4] Louttit's professional involvement with applied psychology apparently displeased his scientific mentor Yerkes, and their relationship was cool. The depression began during Louttit's second year in the islands, adding to the already formidable problems of long-distance job hunting. He was able to secure a one-year teaching appointment at Ohio University for the 1930–1931 academic year.[5]

Louttit's future employment prospects were dim until he heard of an opening as director of the Indiana University Psychological Clinic. There was no question about accepting the offered job, but Louttit was ambivalent about continuing in applied psychology and still hoped to return to animal experimentation.[6] At Indiana,

2 Porteus to Yerkes, 15 July 1925; Yerkes to Porteus, 20 July 1925; Louttit to Yerkes, 22 July 1925; Yerkes to Louttit, 3 August 1925; CML/Box 16/Binder 1-1.

3 C. M. Louttit, "Reproductive Behavior of the Guinea Pig: I. The Normal Mating Behavior," *Journal of Comparative Psychology,* 1927, 7:247–263; "II. The Ontogenesis of the Reproductive Behavior Pattern," idem, 1929, 9:293–304; "III. Modification of the Behavior Pattern," idem, 1929, 9:305–315. He also worked on reflexes with institute member Raymond Dodge: R. Dodge and C. M. Louttit, "Modification of the Pattern of the Guinea Pig's Reflex Response to Noise," *Journal of Comparative Psychology,* 1926, 6: 267–285. C. M. Louttit, "A Bibliography of Bibliographies in Psychology: 1900–1927," *National Research Council Bulletin,* 1928, no. 65, 1–108.

4 Louttit to Harold C. Bingham, 29 December 1928; CML/Box 11/Folder 208.

5 By 1930 Louttit was not listing Yerkes as a reference on his job applications. Yerkes had similar reactions to a number of other students, most of whom were not continuing his type of research program. See Louttit to H. C. Bingham, 7 October 1932; CML/Box 16/ Binder 1932.

6 As Louttit wrote to a friend: "I would like to get someplace where I could go on with some animal work. . . . But the Fates seem to decree otherwise. . . . There it is, back into applied. Still the place seems too good not to take. And there is really no reason why I should drop all other sorts of psychology." Louttit to Donald Adams, 27 May 1931, CML/ Box 16/Binder 1931.

Louttit found a small but thriving clinic, devoted mainly to the study of maladjusted schoolchildren.[7] In addition to directing the daily work of the clinic, he was responsible for graduate training in clinical psychology.

Despite his misgivings about being an applied psychologist, Louttit turned his research and writing skills toward systematizing the field of clinical psychology. In an effort to rationalize the process of clinical diagnosis, he developed a standard twelve-page form to record the patient's personal history and examination results, and successfully marketed the form as the "Indiana Psychodiagnostic Blank."[8] Concerned over the lack of agreement over the definition, scope, and role of clinical psychology, Louttit wrote a systematic textbook, *Clinical Psychology: A Handbook of Children's Behavior Problems,* published in 1936. One of the first textbooks in the field, it reflected the author's interest in producing an empirically based synthesis of useful knowledge within a generally behavioristic theoretical framework. For Louttit, clinical psychology was a field of applied psychology derived from the basic science of psychology as well as from aspects of medicine, education, and sociology.[9]

Louttit also became active in professional affairs during the 1930s. In 1935 he was among the organizers of the Indiana Association of Clinical Psychologists, one of a number of similar state and local associations of applied psychologists begun around the same time to promote nonacademic service roles for psychology. These geographically dispersed groups gave rise to a national organization, the American Association for Applied Psychology (AAAP), in 1937, that provided an alternative to the American Psychological Association's complacent stand on professional issues. Louttit was a leading figure in the AAAP, serving as executive secretary and on various committees; in 1942 he was elected president.[10]

Louttit's slow advancement at Indiana – he was not promoted to associate professor until 1938 – did not keep pace with his increasing professional prominence, and when the possibility of getting involved in military mobilization occurred, he jumped at the chance. His first contact with the military came at a roundtable on "Possible Psychological Contributions in a National Emergency" held at the 1939 annual meeting of the AAAP, where a navy officer mentioned potential interest in

7 C. M. Louttit, "The Indiana University Psychological Clinics," *Psychological Record,* 1937, *1:*449–458.

8 C. M. Louttit, "A Blank for History Taking in Psychological Clinics," *Journal of Applied Psychology,* 1934, *18:*737–748; C. M. Louttit and W. B. Waskom, "The Indiana Psychodiagnostic Blank," *Indiana University Psychological Clinics, Publication Series II,* 1934, no. 7.

9 C. M. Louttit, *Clinical Psychology: A Handbook of Children's Behavior Problems* (New York: Harper, 1936).

10 Among his contributions was the compilation of a biographical directory; see C. M. Louttit, ed., *Directory of Applied Psychologists* (Bloomington, Ind.: AAAP, 1941; 2d ed. 1943).

a psychological unit in the Bureau of Medicine and Surgery. He applied to the navy in June 1940 and in October was commissioned as a lieutenant commander and assigned to the Bureau of Medicine and Surgery.

Louttit's first job entailed working in cooperation with navy psychiatrists to devise screening procedures for navy recruits. Group intelligence tests had been used previously at recruiting stations, but Louttit was able to expand the psychologist's role to include administering individual tests to determine special abilities as well as general fitness for service.[11] But the navy developed no overall system for integrating psychologists into their personnel activities, and Louttit's work was replicated only sporadically at other centers.[12]

Louttit's ambitions to develop a significant role for psychology in the navy were thwarted when he was recalled to Indiana University in August 1941 by his department chair. From that point his navy career proceeded erratically, subject to bureaucratic vagaries. After two months at the university he was ordered to begin work in the Psychological Division of the Office of the Coordinator of Information (renamed later as the Office of Strategic Services). He served there until recalled to active duty in May 1942 to the Training Division of the Navy Bureau of Personnel. His stint included a tour of sea duty on the USS *Iowa* during its shakedown cruise in the summer of 1943. He ended his military career as a commanding officer, first at the Naval Training School in Plattsburgh, New York, and then at the Service School Command, Naval Training Center, Bainbridge, Maryland. Unlike his counterparts in the army and air forces who created and staffed large-scale programs in personnel psychology, Louttit was unsuccessful in creating a similar program in the navy, mainly because he lacked support from high-level military officials. Louttit's military career also hampered his participation in wartime professional affairs, most notably in the amalgamation of the American Psychological Association and the American Association for Applied Psychology, despite the fact that he was serving as president of AAAP in 1943.[13]

The Development of Clinical Psychology

At the start of World War II clinical psychology had not yet achieved a secure place in academic psychology. Defense preparations went hand in hand with continuing

11 C. M. Louttit, "Psychological Work in the United States Navy," *Journal of Consulting Psychology,* 1941, 5:225–227.

12 C. M. Louttit (summary of naval work), 27 July 1944; CML/Box 13/Folder 232.

13 Robert Yerkes wanted Louttit to be secretary of the Intersociety Constitutional Convention that spearheaded the reform of the APA, but Louttit was away at sea. Yerkes to Louttit, 10 April 1943; Louttit to Yerkes, 17 May 1943; CML/Box 12/Folder 225. (See also Chapter 2.) For Louttit's postwar career, see James H. Capshew, "C. M. Louttit (1901–1956): The Career of a Reluctant Clinical Psychologist," unpublished paper delivered at annual meeting of the Cheiron Society in 1986.

efforts to professionalize the field. In the decade of the 1930s psychologists specializing in clinical work sought increased support and recognition for their activities. The field was in a state of intellectual and organization ferment. The social and economic dislocations of the depression, combined with the rise of totalitarian ideologies, helped to focus attention on the role of psychology in reconstructing the social order. Clinical psychologists were anxious to gain professional status in the mental health field, claiming a unique role in diagnosis and treatment. Those employed in nonacademic settings were usually in schools, hospitals, prisons, and other institutions and were typically under the supervision of psychiatrists. Since the turn of the century they had been seeking professional autonomy and had been unable to gain the endorsement of the American Psychological Association for their professionalizing efforts. In 1937 the American Association for Applied Psychology was organized to advance psychology as a practicing profession. Although its membership largely overlapped with that of the APA, the AAAP provided an ideological challenge to the APA's scientific orthodoxy.[14]

Like social psychology, clinical psychology was being codified in textbooks and incorporated into graduate training programs. C. M. Louttit's *Clinical Psychology,* published in 1936, was a landmark effort to define the scope of the field and provide a comprehensive formulation of diagnostic and treatment methods. In 1939 Carl Rogers presented his system in *The Clinical Treatment of the Problem Child.* In an ambitious attempt to survey and evaluate the burgeoning number of psychological tests, Oscar Buros inaugurated the *Mental Measurements Yearbook* series in 1938.[15] With the increasing demand for training in clinical psychology some universities responded with special training programs; others depended on informal student apprenticeships with interested faculty members.[16] While large numbers of the rank and file were occupied in routine testing, leaders in the clinical

14 No satisfactory history of clinical psychology exists. Standard accounts emphasize, but do not analyze, the importance of World War II. See Robert I. Watson, "A Brief History of Clinical Psychology," *Psychological Bulletin,* 1953, *50:*321–346; Virginia S. Sexton, "Clinical Psychology: An Historical Survey," *Genetic Psychology Monographs,* 1965, *72:*401–434; John M. Riesman, *A History of Clinical Psychology* (New York: Irvington, 1976). A recent volume that concentrates on conceptual and methodological issues is Donald K. Freedheim, *History of Psychotherapy: A Century of Change* (Washington, D.C.: American Psychological Association, 1992). See also Donald K. Routh, *Clinical Psychology since 1917: Science, Practice, and Organization* (New York: Plenum Press, 1994).

15 Louttit, *Clinical Psychology;* Carl R. Rogers, *The Clinical Treatment of the Problem Child* (Boston: Houghton Mifflin, 1939); Oscar K. Buros, ed., *The Nineteen Thirty-eight Mental Measurements Yearbook* (New Brunswick, N.J.: Rutgers University Press, 1938).

16 Columbia developed a graduate specialization track in clinical psychology in the mid-1930s. Other notable university centers included Indiana, Iowa, Minnesota, Pennsylvania, and Yale.

field stressed the worth of research into mental illness and advocated a scientific approach borrowed from experimental psychology.

Although clinical psychology was receiving increased professional recognition by the late 1930s, it retained a distinctly subordinate position in the status hierarchy defined by the experimentalist ethos. Efforts to achieve parity for the concerns of practitioners in the American Psychological Association had met with resistance and led to the creation of the American Association for Applied Psychology. The AAAP, which included a Clinical Section, sought legitimacy by adopting restrictive guild standards, making the Ph.D. and postgraduate experience part of its membership requirements. The strategy was to expand into new professional turf by this means, not to set up a rival credentialing program to compete with the existing academic/scientific one. Clinical psychologists had too much invested in the old system to be willing to attempt a completely new alternative.[17]

Control over graduate education was the key obstacle clinicians faced in realizing their professional ambitions. As long as members of the experimental elite headed major departments and continued to train new psychologists as experimentalists, applied psychology would remain a low-status endeavor. In the most sustained attempt to establish a graduate clinical program, Albert T. Poffenberger, author of applied psychology textbooks and chair of the Columbia psychology department, set up a tentative curriculum in 1936 that featured a one-year internship in place of the dissertation. But the proposal was vetoed by the university administration.[18] Eventually, educational reform came from two quarters: academics working in university clinics and clinicians employed in mental hospitals and other institutions. The latter group was particularly anxious to recruit new psychologists as practitioners and researchers into their ranks.

One significant training method that had emerged by the late 1930s was the internship year. An informal system of internships had sprung up at some of the major state mental hospitals and private asylums, providing a link between the academic world of psychology and the realities of clinical practice.[19] David Shakow at the Worcester State Hospital in Massachusetts was prime advocate of this approach, citing his own career as an example of its usefulness.

Shakow (1901–1981) entered Harvard in 1921 as an undergraduate. Because of preparatory work, he had advanced standing as a psychology major. He found coursework in personality development and abnormal psychology interesting and

17 Horace B. English, "Organization of the American Association for Applied Psychology," *Journal of Consulting Psychology,* 1938, 2:7–16.

18 A. T. Poffenberger, "The Training of the Clinical Psychologist," *Journal of Consulting Psychology,* 1938, 2:1–6; William R. Morrow, "The Development of Psychological Internship Training," *Journal of Consulting Psychology,* 1946, 10:165–183.

19 David Shakow, "An Internship Year for Psychologists," *Journal of Consulting Psychology,* 1938, 2:73–76; Morrow, "Internship Training."

took advantage of an opportunity to work at Boston Psychopathic Hospital under Frederic Lyman Wells, who was adapting mental tests for clinical purposes.[20] After he received his bachelor's degree in 1924, he worked for over a year at Worcester State Hospital in order to earn money for graduate school. There he gained more experience in clinical psychology, this time working with another diagnostic test developer, Grace Kent.

When Shakow returned to Harvard for graduate study, he enrolled in the regular doctoral program in experimental psychology. His dissertation research, supervised by Edwin Boring, explored the topic of subliminal perception. Unfortunately, the results were not conclusive enough for Boring, and Shakow was not awarded the doctorate. Financial necessity forced his return to Worcester State Hospital in 1928, where, with his 1927 M.A., he became chief psychologist and director of psychological research. Shakow's slow and difficult progress toward the Ph.D. in experimental psychology had set him more firmly on the road to a career in clinical work.

Over the next dozen years Shakow did extensive research on schizophrenia at the hospital, utilizing both psychometric and experimental approaches.[21] In addition, he established and managed an active internship program for psychologists interested in clinical training. Based on his experience with his own "do-it-yourself" student internships under Kent and Wells, Shakow created a year-long program designed to bring aspiring clinicians into direct contact with patients. By the end of World War II nearly a hundred students had been trained at Worcester, representing a significant portion of the nascent profession of clinical psychology.[22] Finally, in 1942, Harvard awarded Shakow a doctorate in psychology for his dissertation on schizophrenia.

Conference on the Training of Clinical Psychologists

As psychologists continued their civilian war preparations, a conference was held on the training of clinical psychologists in early 1941. Its organizer, Donald B. Lindsley, director of the Psychological and Neurophysiological Laboratories at the Bradley Home in Providence, Rhode Island, proposed to bring together an informal group to study the issue. Lindsley had a strong scientific orientation and was mainly interested in research on psychopathology. C. M. Louttit, executive secretary of the American Association for Applied Psychology, thought the meeting was

---

20 See Frederic L. Wells, *Mental Tests in Clinical Practice* (Yonkers-on-Hudson, N.Y.: World Book, 1927).

21 David Shakow, "The Worcester State Hospital Research on Schizophrenia (1927–1946)," *Journal of Abnormal Psychology,* 1972, *80:*67–110.

22 David Shakow, "An Internship Year for Psychologists (with Special Reference to Psychiatric Hospitals)," *Journal of Consulting Psychology,* 1938, *2:*73–76; idem, "The Worcester Internship Program," *Journal of Consulting Psychology,* 1946, *10:*191–200.

a good idea but was concerned that only half of the fourteen suggested participants were members of the AAAP.[23] Lindsley replied that the idea had developed "spontaneously" and no attempt had been made to be systematic in the selection of participants. He hoped that the informal meeting would at least "get the ball rolling" for consideration of the issues.[24]

Applied psychologist Albert Poffenberger of Columbia served as moderator for the one-day meeting. Carl Rogers's definition of clinical psychology as "the technique and art of applying psychological principles to problems of the individual person for purposes of bringing about a more satisfactory adjustment" was generally agreed upon, and the group divided the functions of the clinical psychologists into diagnosis, treatment, research, and teaching.[25] Stimulated by the meeting, David Shakow of Worcester State Hospital prepared a twenty-four-page proposal outlining a suitable program of professional training in clinical psychology and circulated it to the conference group. Louttit, along with former AAAP president Edgar Doll, director of research at the Vineland Training School, reacted against Shakow's emphasis on psychopathology and its treatment in the psychiatric hospital setting. They advocated a program that provided students with more background in the social sciences and internship training in institutions such as prisons, schools, and social welfare agencies as well as hospitals.[26] Both men were also disturbed by the way the conference overlooked ongoing activities of the AAAP on the issues. After congratulating Lindsley on starting "something of serious importance," Doll went on to chide him, saying:

> the report of the conference itself, however, shows a certain ingenuousness and a lack of familiarity with the antecedents of the past forth [*sic*] years. After all, we are not starting *de novo,* and there is no reason to ignore the past. Since so many clinical psychologists enter the field as novices and make their own way (more power to them!), there is a tendency to sophomoric discovery of what the more sophisticated have known for some time. There is also thus a tendency to start afresh rather than to build on earlier foundations.[27]

23 Lindsley to Louttit, 26 March 1941; Louttit to Lindsley, 29 March 1941; CML/12/223.
24 As it happened, Louttit inadvertently missed the meeting. Lindsley to Louttit, 10 April 1941; Louttit to Lindsley, 6 May 1941; CML/12/223.
25 "Conference on the Training of Clinical Psychologists" (3 May 1941), pp. 1–2; CML/12/223.
26 Louttit to Lindsley, 30 June 1941; Doll to Louttit, 7 July 1941; Doll to Lindsley, 7 July 1941; CML/12/223. Louttit wrote to a sympathetic colleague: "I also can't help but feel that the group as now organized is a little too heavily weighted with people from institutions and with academic people whose interests revolve around factors largely in common with the institution workers. I am afraid clinics, as only two examples, are not too well served." Louttit to E. Kinder, 3 August 1941; CML/12/223.
27 Doll to Lindsley, 7 July 1941; CML/12/223.

Louttit bluntly stated, "I believe this whole problem is one within the AAAP, with the APA subsidiary" and suggested that Poffenberger, as AAAP president, spearhead the creation of a joint committee.[28]

Lindsley backed off a bit, encouraging Louttit and other AAAP officials to take the lead in establishing training standards, including some kind of certification board. He mentioned that Robert Yerkes had written to him about trying to professionalize the entire field of applied psychology, not just clinical work. In August the group held another meeting; discussion centered around Shakow's proposal. In September the AAAP appointed a Committee on Training in Applied Psychology, covering clinical, educational, and industrial psychology. The war impeded the committee's work but, more importantly, radically restructured the situation in applied psychology, giving it increased legitimacy and higher status. Thus the fate of clinical psychology became tied to general professional reform.[29]

Publicity surrounding the proposed merger of the APA and the AAAP, combined with the circulation of a "Proposed Program of Professional Training in Clinical Psychology" to American psychology departments by the AAAP committee and the publication of Shakow's proposal, led to widespread concern over professional issues in psychology.[30] For instance, Robert Sears, the new director of the Iowa Child Welfare Research Station at the University of Iowa, suggested to APA president John Anderson that the APA also appoint a committee on training. Citing his own efforts to develop a joint training program in applied work with the Iowa psychology department, he stated:

> I feel that this is a rather important matter to be undertaken by the APA because it represents the more scientifically minded and academically oriented group of psychologists. The AAAP has, of course, developed a very strong program, and their committees are apparently headed in the right direction. The APA, however, still retains an academic prestige position, and an influence in psychology, particularly on the training side, that no other organization can develop in a short time.[31]

Anderson concurred with Sears and soon appointed an APA Committee on Graduate Training.[32] The APA and AAAP committees, which had some overlapping membership, made Shakow head of a joint Subcommittee on Graduate Internship

28 Louttit to Lindsley, 30 June 1941; Louttit to Lindsley, 18 July 1941; CML/12/223.
29 AAAP, Committee on Training in Applied Psychology, "Proposed Program of Professional Training in Clinical Psychology," *Journal of Consulting Psychology,* 1943, *7:* 23–26.
30 Morrow, "Internship Training"; David Shakow, "The Training of the Clinical Psychologist," *Journal of Consulting Psychology,* 1942, *6:*277–288.
31 Sears to Anderson, 29 January 1943; RRS/2.
32 Anderson to Sears, 16 March 1943; Sears to Anderson, 22 March 1943; RRS/2.

Training in 1944. After the formal merger of the two parent societies, the sub-committee's proposals became the basis for postwar training standards, codified in the "Shakow Report" issued by the APA in 1947.[33]

Clinical Psychology in the Armed Services

Even though a National Research Council advisory committee on psychiatry rec-ommended the use of clinical psychologists by the Army Medical Department as early as June 1941, they were not utilized in the army until the summer of 1942, several months after Pearl Harbor. As an "experiment," six psychologists were commissioned as first lieutenants in the Sanitary Corps. Only three of them held doctorates but each had had practical institutional experience. Assigned to differ-ent general hospitals, their main task was to give and interpret psychological tests. The supervising psychiatrists seemed generally pleased with their work, and one went so far as to complain that his psychological staff was "too small."[34]

This experiment was not followed up until more than a year later, after psychi-atrist William Menninger was appointed chief of the Neuropsychiatry Branch in the Office of the Surgeon General in December 1943. Eclectic and progressive in his views, he took a broad approach to the problem of mental illness and was fa-vorably disposed toward a wide range of approaches to its study and treatment, in-cluding psychoanalysis, clinical psychology, social work, and other specialties. As head of the Menninger Foundation in Kansas, he had integrated a variety of per-spectives in its treatment, teaching, and research programs.[35] Faced with increas-ing numbers of neuropsychiatric casualties, Menninger built upon his experience at the family clinic to reorganize the army's psychiatric program. Among other re-forms, he established a new section of clinical psychology to augment the existing traditional divisions of psychiatry, neurology, and preventative psychiatry (i.e., mental hygiene).[36]

33 APA and AAAP, Committee on Graduate and Professional Training, Subcommittee on Graduate Internship Training, "Subcommittee Report on Graduate Internship Training in Psychology," *Journal of Consulting Psychology*, 1945, 9:243 246; APA, Committee on Training in Clinical Psychology, "Recommended Graduate Training Program in Clinical Psychology," *American Psychologist*, 1947, 2:539–558. See also David Shakow, *Clinical Psychology as Science and Profession: A Forty-Year Odyssey* (Chicago: Aldine, 1969).

34 A. J. Glass and R. J. Bernucci, eds., *Neuropsychiatry in World War II*, vol. 1 (Washing-ton, D.C.: Surgeon General's Office, Department of the Army, 1966); James W. Lay-man, "Utilization of Clinical Psychologists in the General Hospitals of the Army," *Psychological Bulletin*, 1943, 40:212–216, on p. 216.

35 See Lawrence J. Friedman, *Menninger: The Family and the Clinic* (New York: Knopf, 1990).

36 *Neuropsychiatry in World War II*, vol. 1, p. 33.

Before 1944 only a few clinical psychologists had found places in army rehabilitation programs. Menninger's decision opened the door to regular employment
and the integration of clinical psychology into the military establishment. In order
to obtain the services of clinical psychologists, the Surgeon General's Office relied on the aid of the army's chief psychologist, Walter Bingham, in the Adjutant
General's Office. The AGO already employed substantial numbers of psychologists as personnel consultants assigned to various army units. Some of these personnel consultants were clinical psychologists working in induction stations, reception centers, replacement training centers, and in air forces and ground forces
units. When Menninger took over the neuropsychiatry program, one of Bingham's
staff members, Lieutenant Colonel Morton Seidenfeld, served as liaison between
the AGO and the SGO. Seidenfeld had been instrumental in setting up the program
to train illiterate and mildly handicapped army recruits in special training centers
and had had a few years of hospital experience before the war.[37]

By spring, the shortage of psychiatrists to handle a rapidly increasing caseload
was becoming acute, and the SGO advised service command hospitals that clinical
psychologists would be assigned to their neuropsychiatric sections as necessary.[38]
Seidenfeld, chosen to oversee the developing program, was appointed chief clinical psychologist in May 1944. By summer the shortage of psychiatrists had been
quantified: an estimated 300 were needed in addition to the 765 already employed.
But the supply was limited, so the War Department approved the procurement of
175 commissioned clinical psychologists and advertised their availability.[39]

Surveys indicated that immediate needs for clinical psychologists could be filled
from the ranks, so psychologists for the new program were to be selected from
those who were already in the military and who possessed at least a master's degree in applied psychology. Of the 350 officers found to be qualified, 66 already
had hospital assignments. Nearly all the rest were in positions where their psychological expertise was considered essential. But a few were transferred at the start,
and eventually around 130 were obtained in this manner.[40]

Meanwhile, other military service positions were opening up. In May 1944,
Robert Sears became a consultant to the Office of Psychological Personnel and
was asked to make a survey of military clinical psychology. His report was published as part of a special issue on "Clinical Psychology in the Military Services"
in the *Psychological Bulletin*'s "Psychology and the War" section. It revealed the

---

37 Seidenfeld, "Clinical Psychology," in *Neuropsychiatry in World War II,* vol. 1, pp. 570,
    597, 601–602.
38 SGO, memo "Assignment of commissioned clinical psychologists to the neuropsychi-
    atric sections of service command hospitals," 20 April 1944; RRS/4.
39 *Neuropsychiatry in World War II,* vol. 1, pp. 45–46, 570–571; Morton A. Seidenfeld,
    "Clinical Psychology in Army Hospitals," *Psychological Bulletin,* 1944, *41:*510–514.
40 *Neuropsychiatry in World War II,* vol. 1, pp. 570–573.

widespread use of clinical psychology in various programs throughout the services. The impression that clinical psychology was not widely practiced was false, Sears argued, and was based on ambiguous job titles and administrative decentralization. Although military procedures drew from civilian practices, there was a greater emphasis on speed because of large caseloads and other duties and on improving overall group, rather than individual, performance.[41]

The demand for clinical psychologists continued. By August 1944, 213 additional requests had been received. So the army authorized the procurement of another 130 men from the enlisted ranks. With input from the APA and AAAP minimum standards were set. A bachelor's degree in psychology or a related field was necessary for consideration, and further education was highly desirable. Experience in clinical work was also important. Candidates were selected for the training program by the Clinical Psychologist Officer Selection Board, consisting of Seidenfeld, Bingham, and George Evans, chief of the AGO Classification and Replacement Branch.[42]

Training took place at Brooke General Hospital at Fort Sam Houston in Texas, where the Adjutant General's School was located. The original plan, which had called for five training centers around the country, was dropped because of the shortage of psychiatrists to serve as instructors. Six classes of twenty-four trainees were authorized to attend the twenty-two-day course, beginning in October 1944. By the end of the war the training period had been lengthened to thirty-four days, and the course moved to cramped facilities at Camp Lee, Virginia. A total of eight classes completed the course, graduating 281 clinical psychologist officers. Only 3% of the enrollees failed to graduate.[43]

Bingham organized the Advisory Board on Clinical Psychology to help guide the new program. He selected four prominent psychologists with administrative experience in various mental health settings, along with three psychiatrists friendly to professional psychology, to serve.[44] The board held a preliminary meeting in July 1944 to map out their task. The joint SGO-AGO program seemed justified, as

41 Robert R. Sears, "Clinical Psychology in the Military Services," *Psychological Bulletin,* 1944, *41*:502–509.
42 *Neuropsychiatry in World War II,* vol. 1, pp. 574–576.
43 Ibid., pp. 578–581. The original training program was suggested in Seidenfeld to G. Evans and W. Menninger memo "Training of Clinical Psychologists," 22 May 1944; WVB/21. See also C. H. Sievers, "The Current Program of Instruction for Clinical Psychologists at the Adjutant General's School," *Journal of Clinical Psychology,* 1945, *1*:130–133.
44 The original appointees were: Arthur H. Ruggles, M.D., Butler Hospital, Providence, R. I.; Frank Fremont-Smith, M.D., Josiah Macy, Jr. Foundation; Lawrence S. Kubie, M. D., Columbia College of Physicians and Surgeons; Miles Murphy, University of Pennsylvania Psychological Clinic; Frederic L. Wells, Harvard University Hygiene Department; Robert R. Sears, Iowa Child Welfare Research Station; David Rapaport, Menninger Foundation. "Psychology and the War: Notes," *Psychological Bulletin,* 1944, *41*:669.

both army branches were concerned over what to do with soldiers who were "having a tough time of it." The jobs of the psychiatrist and the psychologist were presented as complementary, and closer cooperation was seen to be necessary as the army expanded its rehabilitation and convalescent facilities. Among the major tasks considered appropriate for psychologists was the development of standard diagnostic and evaluation tools, including psychological tests and other measures. Menninger, in particular, emphasized that the clinical psychologist was to be involved primarily in diagnosis rather than therapy.[45]

These bland public pronouncements rested on the common assumption that psychiatrists would remain in control of the neuropsychiatric services and would define the role of the clinical psychologist. Psychiatrists Frank Fremont-Smith and Lawrence Kubie were explicit about this point when they outlined the familiar "psychotherapeutic team" approach, stating "the psychiatrist must have the authority over and the responsibility for whatever is done."[46] Psychologist Sears concurred, emphasizing the special competence of the clinical psychologist in diagnosis and certain types of "educative" therapy. Sears's strategic avoidance of traditional medical turf can also be seen in his advocacy of a research role for army clinical psychologists. He argued that research on psychometric instruments and on the evaluation of psychotherapy would improve the efficiency of military mental health services.[47]

Although army officers routinely attended meetings, the Advisory Board on Clinical Psychology lacked formal policy-making authority. Instead it operated mainly as a forum for the expression of ideas and opinions on mental health topics and served to forge links between psychologists and psychiatrists. Board members seemed well aware of the temporary and ad hoc nature of wartime developments and were more concerned with larger, long-term professional issues. At their first official meeting, in November 1944, the board approved the procurement and training program, advising that it "should now be pushed with all the speed consistent with sound professional standards." Much discussion focused on tests, such as the new Army Individual Test, and the development of a testing manual.[48] Some

45 ABCP, "Proceedings of Preliminary Meeting," 39 July 1944; WVB/21/CCMP, ABCP, Menninger to Bingham, "Memo," 7 August 1944; RRS/4.

46 Fremont-Smith and Kubie, memo "Psychotherapy in the Army," 1 August 1944; RRS/1. The use of an integrated team consisting of psychiatrist, psychologist, and social worker was a popular feature in civilian mental health circles, at least rhetorically if not in actual practice.

47 Sears to Bingham, 6 August 1944; Sears, memo "Research Suggestions for the Consideration of The Adjutant General," 6 August 1944; RRS/4.

48 One member of the board, David Rapaport of the Menninger Foundation, led a notable effort to produce an empirically validated testing manual, of which a condensed version was published during the war with the aid of the Josiah Macy, Jr. Foundation. David Rapaport, with Merton Gill and Roy Schafer, *Diagnostic Psychological Testing: The The-*

participants were concerned over the continued use of the concepts of IQ and mental age, which were misleading. (Bingham had tried to eliminate these terms in all tests developed by the Personnel Research Section.)[49] At the second meeting, in March 1945, continuing shortages of tests were mentioned, along with problems in communicating changing policy directives to neuropsychiatric units. Shortages of clinical psychologists were continuing, and the creation of a new class of workers – clinical psychology assistants – was considered.[50]

The army clinical psychology program also extended to the European theater, but with unimpressive results. Because of problems in implementing the program and complaints by misassigned psychologists, a survey of clinical work was made in March 1945. Many psychologists were found to be employed outside of the neuropsychiatric service, as information and education officers, post exchange officers, and in other positions. Shortages of test materials further hampered the work of psychologists. Attempts were made to remedy the situation by clarifying the role of the clinical psychologist to psychiatrists and administrators, and by holding meetings between psychologists and psychiatrists in hospital centers. But victory in Europe cut these developments short, as personnel were redeployed to other areas. The final statistics of the program reflected its personnel problems: nearly 90% of over six hundred applications for clinical psychologist commissions in Europe were rejected outright, and only fifteen had received the approval of the War Department. As of 1 July 1945 only fifty clinical psychologists, including fifteen newly commissioned, were assigned to general hospitals in Europe.[51]

Conclusion

At war's end, psychologists working in clinical settings had achieved only modest programmatic success, but they were well placed to capitalize on their gains in the postwar era. In short order, thanks to a confluence of social need and professional aspiration, clinical psychology became a full-fledged mental health profession (see Chapter 8).

Within the reformed American Psychological Association, the new status of clinical psychology was symbolized by the election of Carl Rogers as president for 1946. His star rose during the war years as he played the boundary between

ory, Statistical Evaluation, and Diagnostic Application of a Battery of Tests, 2 vols. (Chicago: Yearbook Publishers, 1945–1946). See also the one-volume revised edition edited by Robert R. Holt (New York: International Universities Press, 1968).

49 ABCP, "Minutes of the First Meeting," 6–7 November 1944; WVB/21/CCMP,ABCP.

50 ABCP, "Minutes of the Second Meeting," 29 March 1945; WVB/23/CCMP, ABCP.

51 A. J. Glass, ed., *Neuropsychiatry in World War II,* vol. 2: *Overseas Theaters* (Washington, D.C.: Surgeon General's Office, 1973), pp. 421–425. Apparently, with one or two exceptions, no clinical psychologists worked in the Pacific theater; ibid., p. 449.

scientific and professional psychology. Rogers (1902–1987) had earned a bachelor's degree at the University of Wisconsin in 1924 before attending Union Theological Seminary in New York for two years before transferring to Teachers College of Columbia University to pursue applied psychology. As a graduate fellow he worked at the newly established Institute for Child Guidance, where he came into contact with an "eclectic Freudianism" that contrasted strongly with the emphasis on measurement and statistics at the university. In 1928 Rogers accepted a job at the Rochester Society for the Prevention of Cruelty to Children to perform psychological diagnoses and make treatment recommendations.

Rogers completed his Ph.D. in 1931 with a thesis on a test he devised to measure the personality adjustment of children. He remained at Rochester and eventually became head of a new Rochester Guidance Center in 1939. During the 1930s, as he developed his ideas about child study and therapy, his professional identity became more closely associated with psychiatric social work. His first book, *The Clinical Treatment of the Problem Child,* was published in 1939 and led to an offer to join the psychology faculty at Ohio State University.

At Ohio State, Rogers found that his ideas were considered controversial by psychologists. He continued to develop his nondirective approach and in 1942 published *Counseling and Psychotherapy.* The book articulated his conviction that the effective therapist was a facilitator of psychological change who helped the client (not the "patient") achieve his or her own innate potential for mental health. This approach minimized the status differential between those who gave and those who received assistance and emphasized the interpersonal skills of the therapist. The book included a verbatim transcript of an electronically recorded series of counseling sessions. In keeping with his strong scientific orientation, Rogers hoped that through empirical analysis of the text further dynamics of the therapeutic process would be revealed.[52]

During the war Rogers's developing ideas about the processes involved in therapy received increasing attention. Preferring to remain in his academic role, he served as consultant to civilian agencies as well as the military services. In particular, his interview techniques were widely adopted. By the end of the war, he was a leading figure in clinical psychology, and his election as president of the APA signified the incorporation of the field into mainstream psychology.

52 Carl R. Rogers, *Counseling and Psychotherapy* (Boston: Houghton Mifflin, 1942). See also Carl R. Rogers, "The Processes of Therapy," *Journal of Consulting Psychology,* 1940, *4:*161–164.

7

Engineering Behavior
Applied Experimental Psychology

Personnel work, ranging from the initial selection of soldiers to the rehabilitation of combat casualties, was at the center of psychologists' wartime effort. It absorbed the energies of the largest fraction of the profession and involved a large number of psychologists with no previous experience in test design, construction, and administration. Their work addressed the fundamental task of sorting manpower.

In the great triumvirate of personnel procedures – selection, classification, training – the third task presented a distinct set of challenges. Although existing mental testing techniques were readily adapted for the selection and classification of soldiers, as we have seen, it was less clear how psychology might contribute to the development of effective training procedures. Experimental psychologists, many of whom had previously worked mainly with animals, took up the challenge and applied their knowledge about sensation, perception, and learning to the problems of training people.

Almost from the start of the war, experimental psychologists were involved in applied research on the psychological problems engendered by modern weapons technology. Consciously or not, such weapons were designed with human operators in mind, and the engineers responsible for building them had to rely on some notion of human physical and psychological capacities. Some weapons, such as rifles, had evolved over such long periods that their basic configuration was set and largely unproblematic for the average user. Other technologies, such as range finders and fire-control devices on large shipboard guns, were newer, and the fit between the hardware and its human handler was not always good. Here psychologists found fertile territory for utilizing their expertise, and experimentalists built on the already existing connections with the physical and the life sciences found in such specialties as psychophysics and physiological psychology.

Instead of infiltrating many layers of the military bureaucracy like personnel psychologists, experimentalists tended to remain in quasi-academic settings, supported by government contracts issued to their employing universities. By the middle of the war, the ad hoc and piecemeal approach that had heretofore characterized contracting was regularized to some extent by the creation of the Applied Psychology Panel, under the central Office of Scientific Research and Development (OSRD), headed by Vannevar Bush. Psychologists also began to conceptualize their work in terms of "man–machine systems" and to argue for the indispensable role of "applied experimental psychology" in the design, development, and operation of weapons systems. By the end of the war the value of such work had become clear to military authorities, and they cooperated with psychologists to devise ways to continue to provide federal support in the postwar era.

The Applied Psychology Panel

The Applied Psychology Panel, like the Applied Mathematics Panel, was conceived as an adjunct to the major programs undertaken in the OSRD's regular divisions (e.g., physics, chemistry, explosives, metallurgy, etc.). The idea behind both panels was that their usefulness cut across existing organizational categories; many wartime projects had psychological or mathematical dimensions that could be addressed effectively by this organizational approach. Walter Hunter, an experimental psychologist from Brown and a member of the Emergency Committee in Psychology, was chief of the panel. He was assisted by three technical aides: Charles Bray and John Kennedy of Princeton University and Dael Wolfle from the University of Chicago.[1]

Over the course of the war the Applied Psychology Panel funded twenty major projects, which employed a total of two hundred psychologists. Although the research was usually performed at military field centers around the United States, the contractors were almost always major university psychology departments. Ten universities were involved, along with the Psychological Corporation, the College Entrance Examination Board, and the Yerkes Laboratory of Primate Biology. Under the direction of Hunter's protégé Clarence Graham, Brown University was the largest single contractor. The panel expended a total of $1,500,000, while other units of the National Defense Research Committee spent an additional $2,000,000 on psychological research.[2]

The overarching notion of psychological adjustment achieved new rhetorical forms during World War II, as psychologists confronted the problems of matching

1 Charles W. Bray, *Psychology and Military Proficiency: A History of the Applied Psychology Panel of the National Defense Research Committee* (Princeton, N.J.: Princeton University Press, 1948), pp. 23–25.

2 The universities that received contracts were: Brown, Harvard, Iowa, Pennsylvania, Pennsylvania State, Princeton, Southern California, Stanford, Tufts, and Wisconsin. For a list of major projects see ibid., p. 38.

human capacities to the technologies of modern warfare. Drawing from previous efforts to characterize psychological problems in terms of engineering, the subject was variously defined as engineering psychology, human engineering, biomechanics, psychological problems in equipment design, the human factor in equipment design, applied psychophysics, and psychotechnology.[3] One of the significant concepts that emerged was what Walter Hunter called the "man–machine unit."

Experimental psychologists argued that their knowledge of the human factor – men's aptitudes, skills, and performance limits – was essential to the design and operation of military equipment. As Hunter put it:

> Wars are not fought by machines nor by men alone, but by man–machine units. The machine must be designed for the man, and the man must be selected and trained for the machine. In the construction of tools for war, the engineer recognizes that each individual part must be tested and not be accepted merely on the basis of an offhand opinion. It is not always clearly understood the same care needs to be taken with the man who operates the tool, i.e., with the quality control of the human factor. Efficient human performance depends on a multitude of capacities and abilities which must be analyzed and correlated with the demands of the total job if an efficient man–machine unit is to result.[4]

The rhetoric of engineering proved highly attractive in wartime. It seemed to offer experimentalists, many of whom had confined their prewar work to rats and other laboratory animals, a way to bridge the gap to humans.[5]

Although the ideas and language of such engineering models sounded similar to those of more traditional applied psychologists, there were important differences. They derived more from laboratory research than from mental testing theory and practice. In particular, this engineering approach relied on operational definitions and behaviorist methods that were fashionable among experimental workers.

As he tried to strengthen further the claim of psychology's fundamental importance, Hunter extended the metaphor of the human machine:

> When we speak of tools for warfare, we think of artillery sights, proximity fuses, radar, radio, and atomic bombs. We forget that the brain, eyes, and ears of man are also tools, in fact that they are the indispensable tools of war. Actually, the material equipment of armies and navies is designed solely to

3 Alphonse Chapanis, Wendell R. Garner, and Clifford T. Morgan, *Applied Experimental Psychology* (New York: Wiley, 1949), p. 2.

4 Walter S. Hunter, "Psychology in the War," *American Psychologist,* 1946, *1*:479–492, quoted on p. 479.

5 On earlier examples of the rhetoric of engineering in psychology, see JoAnne Brown, *The Definition of a Profession: The Authority of Metaphor in the History of Intelligence Testing, 1890–1930* (Princeton, N.J.: Princeton University Press, 1992).

extend the effective range of the serviceman's sense organs on the one hand and the striking power of his muscles on the other hand.[6]

Hunter's metaphor resonated with basic ideas about the reflex arc that had arisen in neurophysiology and been incorporated into the foundations of experimental psychology decades earlier. It also drew a tighter connection with engineering practices in contrast to Robert Yerkes's less specific notion of "mental engineering" that evoked the aims more than the methods of traditional engineering disciplines.

Thus the man–machine unit demonstrated how a theoretical rationale was constructed for psychological research and application over the entire spectrum of military technology. In practice, however, institutional support for psychological work on military hardware developed on an ad hoc and loosely coordinated basis. As it happened, psychologists' wartime activities were concentrated in four main areas: vision, fire control, communications, and selection and training.

At first much of the work of the Applied Psychology Panel dealt with improving selection and classification procedures. Specific tests were devised for the selection of radar operators and personnel operating naval guns. The typical approach involved analyzing existing selection procedures and developing standards to measure human performance on the job. In light of the resulting tests of operator aptitude and efficiency, selection procedures were modified. In some cases, such as the selection of night lookouts aboard navy ships, existing tests of visual ability were all proved ineffective, without a suitable alternative being proposed. In the area of classification, the task was to distribute manpower efficiently among an array of special jobs to be performed. Among the major projects undertaken by the panel in 1943 was assisting the navy in the classification of the entire crew of the USS *New Jersey,* the first 45,000-ton battleship in the U.S. fleet. In an effort quickly to launch the vessel, which carried a crew of 2,600 sailors, two-thirds of whom had never been to sea before, the navy turned to psychologists for help. Their job was "to analyze the nature of the ship's many duties, to test and interview each man, and to recommend an assignment for each which he could learn most rapidly and in which he could work most efficiently." The emergency effort was deemed successful by navy officers, who reported fewer problems than usual during the shakedown cruise.[7]

During the latter stages of the war the panel became more involved in developing training procedures and designing equipment. In operating modern military machinery proper training was critical. Using complicated equipment in stressful environments was a difficult task, and developing skilled operators was essential

6 Hunter, "Psychology in the War," p. 481.
7 John L. Kennedy, "Principles and Devices for Military Classification," in Dael Wolfle, ed., *Human Factors in Military Efficiency: Volume 1, Aptitude and Classification* (Washington, D.C.: Office of Scientific Research and Development, 1946), pp. 103–117, quote on p. 104.

to combat effectiveness. The panel supported work in training radar operators, B-29 gunners, winch operators, telephone talkers, and radio operators, among other military specialists. It also helped to develop general procedures that could be more widely applied to military training. As in selection and classification, the determination of valid performance criteria was a prime goal of the research.

Psychologists recognized that selection, classification, and training procedures could be greatly enhanced if psychological factors were taken into account in the initial design of equipment instead of being considered only after it was in operation. Equipment design, however, was the jealously guarded province of engineers. That, combined with the long lead times common in technological innovation, meant that psychologists only made initial progress toward becoming recognized experts in man–machine systems. Research to improve the design of antiaircraft guns, field artillery, and other devices was performed, leading in some cases to specific recommendations for changes in military hardware.

Many of the threads involved in human factors research in the military came together in a programmatic effort to develop "standard operating procedures" through empirical research. Developing such procedures involved systematic study of the interface between the physical characteristics of the equipment and the psychological characteristics of its human operator. The goal was to narrow the margin for human error, which could be achieved through improved training procedures, better equipment design, or both.[8]

The Harvard Psycho-Acoustic Laboratory

The largest university-based program of wartime psychological research took place at the Harvard Psycho-Acoustic Laboratory (PAL). Employing nearly fifty people, including some twenty Ph.D. psychologists, the laboratory dealt with the problems of voice communication in the mechanized cacophony of modern warfare. It was headed by S. S. "Smitty" Stevens (1906–1973), a specialist in auditory psychophysics trained at Harvard under Edwin Boring in the 1930s and retained as a member of the faculty. He proved to be a gifted research administrator as PAL gained major funding from the physics unit (Division 17) of the National Defense Research Committee.

The Psycho-Acoustic Laboratory was set up in tandem with the Harvard Electro-Acoustic Laboratory, headed by physicist Leo L. Beranek. The laboratories made a coordinated attack on the psychological and physical problems of sound control. Both were begun in December 1940 under the auspices of the National Research

8 William E. Kappauf, Jr., "The Development of Standard Operating Procedures," in Dael Wolfle, ed., *Human Factors in Military Efficiency,* vol. 2: *Training and Equipment* (Washington, D.C.: Office of Scientific Research and Development, 1946), pp. 296–302.

Council's Sound Control Committee, chaired by MIT physicist P. M. Morse.[9] Funded by the National Defense Research Committee, the sound control group was formed in response to requests from the army air forces for studies of the effects of noise on aircrew personnel and investigations of methods to reduce noise in military vehicles.

The first major task of the Harvard laboratories was to measure and design sound-absorbing materials for airplanes and other vehicles. Measuring the acoustical properties of existing materials proved to be the most challenging problem; once that was solved, they provided their results and advice to the manufacturers of such materials. Significant progress on improving sound control materials and procedures was made, culminating in a handbook, *Principles of Sound Control in Airplanes.* Soon the research program expanded to include nearly all aspects of voice communication in military vehicles. The PAL became involved in efforts to standardize radio and communication equipment used by the United States Army, the United States Navy, and their British counterparts. The Joint Radio Board of the Joint Aircraft Committee requested information on the design of microphones and headphones for various applications, and the Harvard researchers provided expert advice through the production stage.[10]

In order to test the effects of sustained loud noises on human performance, a sound generator was constructed that could produce an auditory spectrum similar to that of an airplane. The noise could be heard blocks away from the Harvard campus. Volunteers were subjected to a 115 dB sound for seven hours a day over the course of a month and were tested for its effects on psychomotor efficiency through an extensive series of measurements of hearing, vision, motor skills, blood pressure, metabolism, and other variables. Surprisingly, loud noise alone did not significantly affect performance, although temporary hearing loss was experienced. Related experiments attempted to find ways of using loud sounds as a weapon, which proved impractical because of the high energy required to produce them.[11]

9 At a preliminary conference in November 1940, Harvard physiologist Hallowell Davis suggested that research in psychology as well as physics would be needed. Davis, a specialist in audition, had coauthored a textbook with Stevens a few years earlier: S. S. Stevens and H. Davis, *Hearing: Its Psychology and Physiology* (New York: Wiley, 1938). See also Philip M. Morse, *In at the Beginnings: A Physicist's Life* (Cambridge, Mass.: MIT Press, 1977), pp. 159–161.

10 James Phinney Baxter III, *Scientists against Time* (Boston: Little, Brown, 1946), pp. 188–189; P. M. Morse to V. Bush, 11 December 1940; P. M. Morse to K. T. Compton, "Contract Renewal for Sound Control Committee," 11 May 1942; Psycho-Acoustic Laboratory Records/Harvard University Archives/UAV 713.9021/2/NRC, CSC, 1940–42.

11 For a research review and bibliography, see Mark R. Rosenweig and Geraldine Stone, "Wartime Research in Psycho-Acoustics," *Review of Educational Research,* 1948, *18:* 642–654. Stevens provided an engaging popular account of PAL's work in "The Science of Noise," *Atlantic Monthly,* 1946 (July), *178*(1):96–102.

In addition to helping provide specifications for communications equipment, the testing program led to improved training procedures and operating techniques. Factors relating to the intelligibility of speech in military environments were isolated and measured. Articulation tests helped determine how well sound systems transmitted speech, and the role of distortion and interference was examined. Selection tests for operators were devised, and standardized phrases and command forms were developed to reduce mistakes. The laboratory also conducted studies on hearing loss and hearing aids. Commercial hearing aids were evaluated and specifications for improved models drawn up. Pioneering research on the use of auditory signals for instrument flying was performed by PAL in an attempt to relieve the increasing visual load on aircraft pilots. The "FLYBAR" (Flying By Auditory Reference) program led to the development of an automatic annunciator that used synthesized speech to report meter and dial readings.[12]

The institutional style of the Psycho-Acoustic Laboratory was shaped by the personality of its director, S. S. Stevens. Known as "Smitty" to all, he ruled with a firm paternalistic hand. Hardworking and dedicated himself, he expected the same from his subordinates. Although sometimes gruff, he was deeply concerned about the welfare of his laboratory "family" both on and off the job. He expected a full day's work from lab members and could be severe in castigating those who arrived late. But recreation was important too, and Stevens took the lead in organizing social events and outings. His concern extended to every detail of the lab's operation from setting up experiments to office procedures. His able administrative assistant, Geraldine Stone, began her long and close association with Stevens in 1940 (they eventually married, in 1963). Like his mentor Boring, Stevens was zealous in the pursuit of literary excellence and lavished attention on producing clear, well-written reports. Over the course of the war dozens of technical documents were published by the laboratory.[13]

As one of his chief wartime aides later noted, Stevens "was an astute judge of intellectual horseflesh." As a result of his wide contacts among psychologists and acquaintance with electronics specialists Stevens was able to gather together an impressive team of young researchers. The Psycho-Acoustic Laboratory became a major nexus of the wartime psychology network. For example, in 1943 Harvard researcher Clifford Morgan, a fellow instigator of the Psychological Round Table

---

12 Baxter, *Scientists against Time,* p. 190; Rosenweig and Stone, "Wartime Research in Psycho-Acoustics."

13 On Stevens's administrative style, see George A. Miller, "Stanley Smith Stevens," *National Academy of Sciences Biographical Memoirs,* 1975, *47:*428–433. Memos on working hours and office procedures, including a fifteen-page publication manual, can be found in PAL/UAV 713.9035/1. Well over a hundred reports were produced; see "Check-List of Reports and Informal Communications Issued by the Psycho-Acoustic Laboratory"; PAL/UAV 713.9021/2/NDRC Section 17.3, Bi-Monthly Progress Reports.

a few years earlier, was appointed OSRD technical aide to the laboratory. His wife, Jane Morgan, was in turn employed by Boring and then by the Office of Psychological Personnel in Washington, where she could refer prospects to the PAL. In such fashion it developed into a productive organization, with a host of promising graduate students and bright postdocs who were permanently imprinted on its interdisciplinary, problem-centered approach to psychological research.[14]

Stevens also proved adept at maintaining a high level of support for the work of the PAL from its sponsors and clients. The entire Harvard Sound Control Project (including the Electro-Acoustic Laboratory) received nearly $2 million from the National Defense Research Committee during the war, making it the largest single recipient of federal aid for psychological research. Thanks to Stevens's promotional efforts, the laboratory and its work enjoyed high visibility. Part of its mystique was due to its specialized facilities, including a loud noise generator and an anechoic chamber touted as the "world's quietest room." In addition to its scientific utility, the sound generator served as a useful public relations gimmick. Leading the Harvard Board of Overseers on a tour of the laboratory, Stevens sought "to demonstrate that modern warfare is getting noisier and noisier. The ancient art of the war-cry has been drowned out by a superior din, and the modern problem is not so much how to blitz the enemy with a racket, as how to keep coordinated an intricate fighting unit in the face of thunderous accompaniment." Visitors could hardly remain unimpressed by the problems of voice communication on aircraft as Stevens instructed them to cover their ears when he turned on the noisemaker.[15] Equally impressive was the anechoic chamber, with its catwalk suspended in the midst of thousands of sound-absorbing Fiberglas wedges padding the walls. Stevens described its dramatic effect on the first-time visitor: "So oppressive is the sudden quelling of all the familiar noises which in our normal lives beat unheeded on our eardrums that we obey an instinctive urge to swallow and rub our fingers in our ears in an attempt to dislodge some nonexistent obstruction." Drawing an analogy to a widely publicized tool of the nuclear physicist – the cyclotron – Stevens pointed out the necessity of such special environments for the conducting of basic research in acoustics.[16]

14 First quote from Miller, "Stevens," p. 430. J. Morgan to Stevens, 19 August 1943; PAL/ UAV 713.9035/1/NRC,OPP. Among the psychologists associated with the PAL were J. C. R. Licklider, George A. Miller, Edwin B. Newman, and John Volkmann; see William R. Woodward, "Stevens, Stanley Smith," in Frederic L. Holmes, ed., *Dictionary of Scientific Biography,* vol. 18, Suppl. II (New York: Scribner's, 1990), pp. 869–875.

15 Although PAL research was classified, its presence on campus was no secret, especially because the noises the lab generated could be heard several blocks away. Stevens to I. Stewart, 11 June 1943; PAL/UAV 713.9021/1/OSRD re Security. Stevens ("Talk to Board of Overseers"), 11 May 1942; PAL/UAV 713.9021/2/Harvard University, Board of Overseers.

16 Stevens, "Science of Noise," p. 96.

Early in the war the Harvard Sound Control Project came under criticism for its independent style from Irvin Stewart, the executive secretary of the Office of Scientific Research and Development and Vannevar Bush's chief administrative assistant. Both the Psycho-Acoustic Laboratory and the Electro-Acoustic Laboratory worked closely with the armed services actually using the communications equipment and soundproofing materials and with the companies manufacturing it, and promoted the free exchange of information and reports among all interested parties. Under government statutes, however, the information produced through contract research and development was legally the property of the Office of Scientific Research and Development. Stewart put pressure on the Harvard project to conform to regulations about the distribution of reports by castigating its sponsor, the National Research Council.[17]

A year and a half later top federal science officials became critical of the amount of testing the laboratories conducted for the armed forces and labeled the project as "not highly essential." This prompted a strong rebuttal from Morse. He admitted that the work under Stevens and Beranek had differed from other projects, making it difficult to fit it into "the usual straight jacket of NDRC procedure." But the difference – its close relation to the armed forces and equipment manufacturers through all phases of research, development, and production – was a valuable one. Most other NDRC projects entailed the development of a specified piece of military equipment, and once the model or prototype was delivered, project personnel usually had minimal involvement with production and implementation. Morse cited the development of tests and specifications for soundproofing materials and communications equipment that the group had produced and pointed out the advantages of setting such standards first and then measuring and modifying existing materials instead of starting from scratch. In the case of aircraft telephones, Morse estimated that the Harvard approach saved at least six months in getting effective equipment to operators in the field. In conclusion he stressed that the mission of the Sound Control Project was to develop universal methods and standards, not conduct routine testing for the army or navy.[18]

In fact, the relationship between the project and its military clients was not always as smooth as the NDRC thought. For instance, when the Psycho-Acoustic Laboratory tried to introduce one of its most simple and useful innovations, improved earplugs, it encountered some resistance from segments of the military. After performing extensive tests on existing earplug designs, the PAL came up with its own model, dubbing them "Ear Wardens." In seeking to get them adopted by the army ground forces, one of the staff visited its procurement offices. Despite the approval of the earplug by the army's Field Artillery Board, some officers were

---

17 Stewart to A. L. Barrows (NRC Secretary), 21 January 1942; PAL/UAV 713.9021/2/ NDRC C-5.
18 Morse to Harvey Fletcher, 24 July 1943; PAL/UAV 713.9021/2/NDRC Sec. 17.3.

reluctant to switch from the traditional cotton plug. But most became convinced after a demonstration and discussion of the Ear Warden's advantages. One colonel, however, was entirely opposed; the PAL representative reported: "He does not care whether the NDRC development is the most scientific earplug ever made. Cotton has been used satisfactorily for years. . . . even if the men had the plug, they would not use it, because there is no need for it." Such objections eventually subsided and substantial numbers of the Ear Wardens were ordered.[19]

The wartime productivity of the Psycho-Acoustic Laboratory led to its continuation after 1945 as federal contract funds flowed unabated. Stevens's own career flourished. He was granted tenure and promoted in 1944, after being told a few years earlier by President Conant that he could expect to remain an assistant professor permanently. The cadre of experimental psychologists he had assembled helped to revitalize the Harvard program.[20]

B. F. Skinner and Project Pigeon

Psychologists displayed ingenuity and originality in making their science relevant to national needs during the war. They temporarily abandoned their reliance on the animal laboratory as a main source of professional status and broadly defined their expertise as relevant to practically any activity involving human behavior, thereby creating a flexible, all-purpose rationale for their work. Problems of military personnel, civilian morale, psychological warfare, and man–machine engineering were readily considered under this rubric. The wholesale shift toward human subjects was so complete that one exception provided a singular contrast.[21]

During the years surrounding the Second World War, B. F. Skinner was a bright young star among the junior faculty of the University of Minnesota but little known outside the circles of psychology researchers. He was one of a number of neobehaviorists who were seeking to place the behaviorist program espoused by John B. Watson on a sound scientific footing through systematic experimental research, mainly on the white rat. Skinner's first book, *The Behavior of Organisms: An Experimental Analysis* (1938), summarized nearly a decade of his laboratory work conducted at Harvard and provided the basis for his scientific reputation.

19 E. S. Russell, "Memorandum," 15 August 1944; Russell, "Memorandum," 5 February 1945; PAL/UAV 713.9021/5/USA, AGF, Army War College. Baxter, *Scientists against Time*, p. 191.
20 Stevens to Paul Buck (Harvard Dean of Arts and Sciences), 8 June 1945; Stevens, "A Proposal for Psycho-Acoustic Research in the Post-War Period," 21 May 1945; PAL/ UAV 713.9021/1/Postwar. S. S. Stevens, in *History of Psychology in Autobiography*, vol. 6 (Englewood Cliffs, N.J.: Prentice-Hall, 1974), pp. 393–420, on p. 414.
21 The following is derived from James H. Capshew, "Engineering Behavior: World War II, Project Pigeon, and the Conditioning of B. F. Skinner," *Technology and Culture*, 1993, *34*:835–857.

Resolutely focused on experimental research but intrigued by the possibilities of applied psychology, Skinner confined his work to the laboratory until the war came along. Prompted by a mixture of patriotic sentiment and intellectual curiosity, he began to ponder the idea of training pigeons to function as missile-guidance systems. After some preliminary research, Skinner presented his unorthodox approach to government authorities and received a $25,000 contract from the guided missile group (Division 5) of the National Defense Research Committee in 1943. The funds were channeled through the General Mills Company in Minneapolis, because of their engineering expertise.

In the wartime context, Skinner's work on "Project Pigeon" was unusual but not outlandish, given the primitive state of guided missile technology. Servomechanisms were only beginning to be developed, and inertial guidance systems had yet to be invented. In dealing with the engineers who managed the contract, Skinner learned to downplay the biological nature of the bird and characterize its behavior in mechanical terms. His conditioning techniques, derived directly from his laboratory research, enabled him to train the birds to perform reliably under stressful conditions similar to those they would encounter in combat.

Essentially, the pigeons were trained to function as kamikaze pilots. Strapped into the nose cone of the missile, they would view the target (for example, a bridge) through a transparent lens and peck at its image. Their pecking movements would provide signals to the controls of the missile and keep it on target until the moment of impact. Skinner and his co-workers, who included Minnesota graduate students William K. Estes, Keller Breland, and Marian Breland, constructed a flight simulator in a space provided by the General Mills Company in an old flour mill. After a year of work, funds ran out just as the project was on the verge of field tests. Skinner tried to persuade the NDRC engineers that the device would work, but they remained skeptical.

Traveling to Washington in March 1944 for a last-ditch plea for more money, Skinner demonstrated the device in front of NDRC officials. He described the scene:

> The translucent screen was flooded with so much light that the target was barely visible, and the peering scientists offered conditions much more unfamiliar and threatening than those likely to be encountered in a missile. In spite of this the pigeon behaved perfectly, pecking steadily and energetically at the image of the target as it moved about. . . . It was a perfect performance, but it had just the wrong effect. One can talk about phase lag in pursuit behavior and discuss mathematical predictions of hunting without reflecting too closely upon what is inside the black box. But the spectacle of a living pigeon carrying out its assignment, no matter how beautifully, simply reminded the committee of how utterly fantastic our proposal was. I will not say that the meeting was marked by unrestrained merriment, for

the merriment was restrained. But it was there, and it was obvious that our case was lost.[22]

After discussion of the project, which revealed a lack of technical supervision on the part of the NDRC, the request for additional funds was denied.

Project Pigeon was officially over. But Skinner had learned much from his initial foray into applied experimental psychology, refocusing his work toward the possibilities of behavioral engineering. During the 1944–1945 academic year he accepted a Guggenheim Fellowship to prepare a monograph on verbal behavior. As he worked on the book during the last year of the war, he explored other forms of behavioral technology. Faced with the challenges of raising a second baby daughter, Skinner drew on his manual skills and invented the "baby-tender," a futuristic climate-controlled crib designed to promote the physical and psychological health of infants. Featured in the *Ladies' Home Journal* shortly following the war, the device was later marketed commercially, with little success, as the "Aircrib." In the summer of 1945 Skinner drafted the manuscript that would be published three years later as *Walden Two*. The book was Skinner's attempt to conceive a utopian human society based on the principles of reinforcement that he had gleaned from his laboratory research on animal behavior.

Skinner dramatized the consequences of his wartime research in his autobiography, published thirty-five years later, saying: "Project Pigeon was discouraging. Our work with pigeons was beautifully reinforced, but all our efforts with the scientists came to nothing. . . . [However,] the research that I had described in *The Behavior of Organisms* appeared in a new light. It was no longer merely an experimental analysis. It had given rise to a technology."[23] Project Pigeon became the opening wedge in what evolved into a campaign for behavioral engineering. The war provided Skinner with an opportunity to redefine the disparate problems associated with guiding a missile, raising a baby, and managing a society in terms of a common behavioral framework and to propose solutions based on techniques derived from the psychological laboratory. His wartime mechanical devices and literary constructions were designed with the same goals as his experimental apparatus: to control behavior predictably. Although Skinner's inventions received mixed reviews, he became convinced that behaviorism offered scientific methods equally applicable outside and inside the experimental workplace. Faced with new contingencies, Skinner changed his behavior as a scientist during the war and began to discover how the laboratory could provide significant leverage in the wider realm of human affairs. Although his experience was undeniably idiosyncratic, it reflected broader trends in American psychology that reached their fullest expression in the environment of World War II, with its overriding emphasis on military utility and the virtues of order, control, and effectiveness.[24]

22 Ibid., p. 852.   23 Ibid., p. 856.   24 Ibid., pp. 836, 857.

Interlude III

By 1944, only a decade after he led psychology at Harvard to independent departmental status, Edwin Boring faced a serious challenge to the status quo he had helped to create. Differences between what Boring called the "sociotropes" and the "biotropes" in the department were threatening to split the faculty into warring camps. The biotropes included Boring and his allies, who emphasized the more traditional and experimental parts of psychology, such as sensation, perception, psychophysics, and psychophysiology. In contrast, the sociotropes were more oriented toward the social sciences in their studies of social, clinical, and personality psychology. At Harvard they included Gordon Allport, Henry Murray, and their associates.

The wartime salience of sociotropic psychology emboldened Murray and Allport to join forces with their colleagues in anthropology and sociology to propose a new department that would focus on social behavior. They argued that existing institutional arrangements favored traditional experimental psychology and would not allow them to pursue their ambitious goal of a unified science of human personality, society, and culture.[1] Although Boring had some sympathy for the sociotropic point of view, he was unwilling to make fundamental changes in the psychology department's program. That reluctance, coupled with efforts of sociologist Talcott Parsons and other powerful Harvard faculty members, led to the creation of the separate Department of Social Relations in 1945. Chaired by Parsons and with Samuel Stouffer directing the laboratory, it brought together sociologists, anthropologists, and psychologists interested in personality and social behavior.[2]

1 On the academic politics in the creation of the new department see: Rodney G. Triplet, "Henry A. Murray and the Harvard Psychological Clinic, 1926–1938: A Struggle to Expand the Disciplinary Boundaries of Academic Psychology" (Ph.D. diss., University of New Hampshire, 1983).
2 Talcott Parsons, *Department and Laboratory of Social Relations, Harvard University: The First Decade, 1946–1956* (Cambridge, Mass.: Harvard University, 1957).

Boring jokingly referred to himself as a member of the department of "unsocial psychology," and took the opportunity to reassert the Department of Psychology's identity as a bastion of experimentalism. One outcome of the fission between the biotropes and sociotropes at Harvard was the construction of a new Psychological Laboratory in the basement of Memorial Hall in 1946. Some of the space had already been refurbished during the war for the Psycho-Acoustic Laboratory, headed by Boring's protégé, psychophysicist S. S. Stevens. The rest was finished soon afterward under the guidance of Stevens, who was Boring's heir apparent as chair of the department. Boring was immensely pleased with the new quarters, where he could sit at Münsterberg's refinished desk and preside over the scene with a sense of patriarchal pride and power.[3] So long as the confirmed biotrope had a home, he could tolerate the growing number of sociotropes in the vicinity.

By the late 1940s Boring had coupled his biotropic/sociotropic distinction to another bifurcation, that between "basic" and "professional" psychology. He classified different specialties within a 2 × 2 table:

	Biotropic	Sociotropic
Basic	Experimental	Personality
	Physiological	Social
Professional	Applied psychophysics	Clinical
	Human engineering	Industrial
		Personnel

The table neatly captured Boring's pecking order, from basic biotropic to professional sociotropic. As applied psychology continued to grow, he expressed concern about "the threat of expanding clinical psychology to the basic values of experimental psychology."[4] After all, if psychology were to dilute its scientific basis too much, it would lose whatever warrant it had for effective applications.

Boring's transition from staunch opponent into grudging advocate of the professionalization of sociotropic psychology was also refracted in his revised edition of *A History of Experimental Psychology,* published in 1950. Both the author and the book had changed significantly since 1929. Now nearing retirement age, Boring was at the height of his intellectual powers and professional influence. Much of the text of the now classic first edition had been incorporated into the new edition, but it was placed within a radically different historical framework. Boring had moved away from his earlier reliance on the great-man theory of history by juxtaposing it to the influence of the Zeitgeist. The Zeitgeist consisted of the "habits of thought that pertain to the culture of any region and period" and was considered

---

3 Edwin G. Boring, *Psychologist at Large* (New York: Basic Books, 1961), p. 66; S. S. Stevens and E. G. Boring, "The New Harvard Psychological Laboratories," *American Psychologist,* 1947, 2:239–243.
4 E. G. Boring to R. R. Sears, 19 March 1948; JGM/Box 1947–49/Misc. MSS prior to 1955.

primarily to be a limiting factor for the creative efforts of individual scientists.[5] This provided a way to explain why certain innovations did not occur; in common parlance, the time was simply not "ripe."

By placing a personalistic theory of history in dynamic tension with a naturalistic approach, Boring could accommodate the creativity expressed by individual scientists while at the same time accounting for resistance to intellectual change. The great-man approach privileged individual free will and the process of discovery; the Zeitgeist stressed the impersonal and inexorable nature of conceptual change. In his words, "a psychologist's history of psychology is, therefore, at least in aspiration, a dynamic or social psychology, trying to see not only what men did and what they did not do, but also why they did it or why, at the time, they could not do it."[6]

Invoking the Zeitgeist as a causal factor in historical change allowed Boring to submerge the individual self (his own included) in the flowing stream of history. It provided a comforting determinism and a warrant for people to go with the flow of events and take a dispassionate view about their own ability to change the course of history. No longer would the burden of intellectual progress rest solely on the shoulders of scientific giants. In contrast to his earlier position, great men were seen as neither the causes nor the symptoms of scientific progress; they were its agents instead. Boring used the Zeitgeist as a residual category, applying it whenever individualistic explanations did not suffice.[7]

Although the second edition of Boring's *History* retained the same programmatic thrust as the first edition, its coverage of nonexperimental and applied psychology was greatly expanded. One indication of this was that Sigmund Freud merited his own heading; he had been mentioned only in passing in the previous edition. Now, as an agent of the Zeitgeist, Freud was a candidate for great-man status.[8] Mental tests also received more attention. In the first edition Boring had briefly reviewed developments up to the turn of the century, concluding that "in America the mental tests began in the psychological laboratory and then wandered away from it."[9] In the second edition he was still uncertain that the history of testing "belonged" in

---

5 Edwin G. Boring, *A History of Experimental Psychology*, 2d ed. (New York: Appleton-Century-Crofts, 1950), p. 3.

6 Ibid.

7 For a critique of Boring's use of the Zeitgeist concept, see Dorothy Ross, "The 'Zeitgeist' and American Psychology," *Journal of the History of the Behavioral Sciences*, 1969, *5:* 256–262.

8 On Freud and psychoanalysis, see Boring, *History of Experimental Psychology* (1950), pp. 706–714; in the concluding chapter he assesses Freud: "Psychologists long refused him admission to their numbers, yet now he is seen as the greatest originator of all, the agent of the *Zeitgeist* who accomplished the invasion of psychology by the principle of the unconscious process" (p. 743).

9 Boring, *History of Experimental Psychology* (1929), p. 549.

the history of experimental psychology, but he nonetheless ascribed a central role to such work in the development of the discipline. During the 1940s, the widespread military use of tests, combined with the rise of statistical methods in psychology more generally, had encouraged a rapprochement between experimental psychology and mental testing.[10]

Boring himself had undergone a sea change, personally and professionally, during World War II. Imbued with a powerful mixture of patriotic duty and professional pride, Boring launched into war work early and enthusiastically. Like others of his generation, he could draw on his experiences in the First World War for guidance. But this war was different for Boring and his colleagues. Longer in duration and more absorbent of the energies of psychologists, it transformed the profession in unanticipated and unprecedented ways as applied psychology took on a new significance. During the war psychologists forged an alliance for scientific professionalism among themselves, ideologically as well as organizationally, that promoted the practical utility of their knowledge and their role as experts on human behavior.

Boring willingly laid his scientific conscience aside for the duration and became a key player in several important wartime projects that furthered this technocratic vision of psychology. In the process of adapting to what he considered a temporary situation, he found his outlook on the field permanently altered. He finally accepted as inevitable the responsiveness of psychologists to social demands for usefulness. The history of the wartime mobilization of psychology had clearly demonstrated that the fortunes of the field as an intellectual enterprise were inextricably linked to its ability to attract support on the basis of its perceived utility. Indeed, the alliance for scientific professionalism rested on the symbiosis between scientific research and professional practice that permitted psychologists to march together despite their epistemological, methodological, and political differences.

As Boring crafted the second edition of *A History of Experimental Psychology* in the years immediately following the war, his changed attitudes toward the profession were reflected in his use of the Zeitgeist theory of history. With this concept, he could put not only his wartime work but his entire career in context, viewing it as his response to the irresistible forces of social and cultural change. It also allowed him to come to terms with the increasingly applied direction of psychology, which could be portrayed as a concomitant or unavoidable by-product of disciplinary development that had little necessary connection with the scientific advances achieved by great men.

10 Boring, *History of Experimental Psychology* (1950), pp. 570–578.

8

A New Order
Postwar Support for Psychology

In late November 1945 a group of about 150 American psychologists gathered near Washington, D.C., to discuss the impact of World War II on their work. Navy Captain John G. Jenkins, chair of the psychology department at the University of Maryland, reflected on the amazing transformation the war had wrought on the profession. In the space of five years military attitudes had changed dramatically. At the start of the war psychologists entered the services "under the conviction that things were so bad that any available magic should be tried, even psychology." A few short years later they were "going out the big front door" with the band playing "Hail to the Psyche."[1] As evidence of wartime professional success, Jenkins cited the establishment of permanent divisions of psychology within the military services.

On a more somber note, Jenkins mused about how the atomic bomb seemed to render such professional self-congratulations beside the point. The use of nuclear weapons to destroy Hiroshima and Nagasaki four months earlier had also "destroyed much of the basic framework of the social world we inhabited before the war," making it impossible for psychologists to return to the academic world they had left a few years earlier. Now psychologists, along with scientists and citizens everywhere, were caught up in the challenges of a new era of global insecurity.[2]

In projecting psychology's ambiguous but exciting future, Jenkins articulated a developmental model that divided the history of the science into three phases. Up to the 1920s psychologists pledged allegiance to local loyalties defined on the basis of particular schools of thought. Rivalries among competing departments were strengthened through geographical isolation and institutional inertia. One joined a

1 John G. Jenkins, "New Opportunities and New Responsibilities for the Psychologist," in George A. Kelly, ed., *New Methods in Applied Psychology* (College Park: University of Maryland, 1947), pp. 3–13, quoted on p. 1.
2 Ibid., p. 2.

psychological "team" and "thenceforth fought lustily to show that your team was right." The second phase was characterized by a broadened sense of loyalty to the profession as a whole. Beginning in the 1920s, the schools lost their influence as their perspectives and findings were increasingly absorbed into the scientific mainstream. The Second World War demonstrated the reality of this phase, Jenkins suggested, as it proved that psychologists from different backgrounds were able to work together easily, united by their common interest in sound scientific reasoning. Now psychology was on the threshold of a new stage in its development that emphasized social responsibility. In response to outside pressures for relevance, psychologists "shall now have to ask not merely whether a result has *statistical* significance but also whether it has *social* significance."[3]

Jenkins embraced this vision of psychology as a science responsive to social demands and sketched its implications for his colleagues:

> Social responsibility does, of course, include a willingness to meet and deal with social issues. But it goes far beyond that. Social responsibility for the research psychologist touches his professional life at every point. It determines what problems he shall select for his attack. It determines, to a considerable extent, what shall be accepted as methodologically respectable methods of attack upon these problems. It also determines, to some extent, how he shall interpret his findings and how and where he shall publish his interpretation.[4]

The veteran psychologist urged his colleagues to focus on "the less well-behaved aspects of human behavior," such as personality, that were messy and difficult but undeniably important.[5] Social relevance was to take precedence over methodological rigor. Psychologists could no longer afford to pursue narrow and self-contained scientific goals, because to turn away from the tasks of social reconstruction and cultural redemption ensuing from the most destructive war in human history would be professionally irresponsible.

Despite its rousing rhetoric, Jenkins's message was hardly radical. It expressed a view of psychology's future that had become commonplace in the patriotic environment of the war. The psychology community, responding to the demands of the national emergency, articulated an ideology that stressed the themes of professional unity, scientific rigor, and social utility. What was different was the postwar context, in which new resources would become available for psychologists to pursue their technoscientific agenda with unprecedented vigor.

The Postwar Environment

During and immediately after the war, psychology strengthened the scientific basis of its cognitive authority and expanded its professional domain. This was due

3 Ibid., p. 5.   4 Ibid.   5 Ibid., p. 7.

partly to the efforts of psychologists, and partly to a general climate that was favorable to science. Perhaps more important, however, were the institutional mechanisms that enabled psychologists to pursue their scientific and professional agendas simultaneously. Psychology departments, modes of communication, the American Psychological Association – all were revamped in accordance with a new sense of unity and solidarity articulated among psychologists during the war. The breakdown of traditional boundaries dividing psychologists – academic and applied, younger and older, liberal and conservative – in the context of wartime mobilization led to a reconsideration and reformulation of the commonalities that bound them together. Education, voluntary association, publications, and other aspects of collective endeavor were remolded to emphasize that diversity was a virtue, that the laboratory and the clinic were symbiotic, that psychology was a capacious science with virtually unlimited possibilities for application.

The Maryland conference was sponsored by the Division of Military Psychology, one of the charter divisions of the reformed American Psychological Association. In the APA reorganization, psychologists had expressed their solid if limited support for such a division. Relatively few psychologists made it their primary specialty identification, but many listed it among their subsidiary interests, suggesting perhaps that psychologists viewed their wartime concerns as a temporary expedient and preferred to identify with more traditional substantive fields.[6]

The revamped American Psychological Association embodied the hopes of U.S. psychologists for an effective vehicle to pursue their newly conjoined scientific and professional aspirations. Officially inaugurated on 6 September 1945, less than a month following the surrender of Japan, the new APA got off to a flying start under the leadership of Dael Wolfle, the first executive secretary.

Not quite forty years old, Wolfle had been a junior faculty member at the University of Chicago before the war. Interested in test development and human learning, he was a member of the Psychometric Society and had served as its representative on the Emergency Committee in Psychology during the war. In 1944 he joined the staff of the Applied Psychology Panel. At war's end he played a major role in bringing together the panel's two-volume summary technical report on *Human Factors in Military Proficiency*. In addition to editing both volumes, he wrote nearly half of the forty chapters.[7]

Wolfle was in charge of the APA's new central office in Washington, D.C. Organized as a clearinghouse and information center, the office fulfilled a number of functions. It served as the publisher for the APA's stable of scientific journals as well as the new journal of professional affairs, *The American Psychologist*. It was

6 Ernest R. Hilgard, "Psychologists' Preferences for Divisions under the Proposed APA By-laws," *Psychological Bulletin,* 1945, *42*:20–26.
7 Dael Wolfle, ed., *Human Factors in Military Efficiency,* 2 vols. (Washington, D.C.: Office of Scientific Research and Development, 1946).

responsible for arranging the annual convention, which was growing into a large gathering. Through its committees and boards it pursued the aspirations expressed in the APA's expanded charter "to advance psychology as a science, as a profession, and as a means of promoting human welfare." Perhaps most important of all it was a visible sign of a unified profession, located near the center of political power, marching confidently into the postwar future.

In January 1946 the *American Psychologist* began monthly publication. Edited by Wolfle, it provided a forum for discussion of policy issues, trends in research and training, and other matters of interest to the profession. In addition it served as a news center, disseminating information on people and events in the world of psychology. For outsiders as well as insiders it was an indispensible source.

Nominally a bureaucrat, Wolfle was a prime example of the new breed of technoscientific professionals who had come to power during the war. By 1945, he was wise in the ways of Washington and had a pragmatic sense of scientific professionalism, and he proved to be an adept manager of the affairs of the American Psychological Association as it became a major reference point for the psychology community and an important means to exert professional influence.

The APA grew rapidly in the postwar decade. Adding an average of 1,000 new members a year, it reached a total of 14,000 in 1956, the year the architect of wartime reform Robert Yerkes died. Boring, Yerkes's erstwhile lieutenant, was ever mindful of the compromises involved in the reorganization. In 1949 he reminded the psychology community that

> whatever happens to us, APA is going to remain a huge organism with two heads, a professional and a scientific. That is what we wanted from the start and what we now have. Let us not forget the advantages it has given us. The two heads are on the same end of the animal, and, if either of them thinks the other is aiming in the wrong direction, let it look around to see where its tail had come from in 1945.[8]

The Changing Demography of APA Leaders

The changes wrought by the wartime reformation of the American Psychological Association proved to be dramatic and enduring. Its inclusive and expandable structure could accommodate substantial growth, as well as provide an organizational center of gravity for heterogeneous disciplinary interests. In addition, by reclaiming its place as the indisputably "main" national organization of psychologists, membership provided an unambiguous indicator of "belonging" to the profession. Although the APA presidency retained its important symbolic function,

8 Edwin G. Boring, "Policy and Plans of the APA: V. Basic Principles," *American Psychologist,* 1949, *4:*531–532.

the office became an exhausting job as well as a high honor as the organization grew.

The presidential cohort provides a convenient index to changing disciplinary interests and institutional patterns. The group that held office between 1920 and 1940 reflected the dominance of neobehavioristic learning theory. The bulk of the presidents represented specialties close to the traditional experimental core of the discipline. There were, however, a few identified with the burgeoning area of mental testing and the development of psychometric methods. At the end of this period, the 1938 election of Gordon Allport as the first APA president identified with personality theory and social psychology signaled the growing acceptance of such fields in the discipline. The 1940–1960 cohort, in contrast, displayed the broadening of psychological research and practice across a spectrum of fields, including social, clinical, and personality. (In Boring's parlance, they tended to be more "sociotropic" than their "biotropic" predecessors.) The election of Carl Rogers in 1946 as the first clinical psychologist to hold the office symbolized the incorporation of clinical work into the mainstream of psychology.[9]

Further analysis of APA leadership reveals changes in the demography of the psychology establishment. Of the twenty presidents who served between 1920 and 1939, 15 (or 75% of the total) had obtained their doctorates from four universities: Chicago, Columbia, Cornell, and Harvard. These four institutions had been among the leading Ph.D. producers in psychology since the turn of the century. The same institutions were also among the leading employers of this group of APA presidents at the time they held office, although employment locations exhibited less concentration than doctoral training sites for the entire group. More striking is the overwhelming preponderance of private eastern universities in the employment of prewar APA presidents. Fully 90% of the group worked at private universities and colleges; over two-thirds were located in the East, with the remainder split between the Midwest and the West Coast.[10]

For the period from 1940 to 1960, there are some interesting continuities as well as dramatic shifts from the prewar demographic situation. Over those two decades the doctoral origins of APA presidents remained concentrated in a few institutions. Fifteen of the group (75%) obtained their Ph.D.s at five institutions: Columbia, Cornell, Harvard, Stanford, and Yale. The first three had already figured prominently in the training of future APA leaders, whereas Stanford and Yale were newcomers to this list, reflecting their rise as important centers in the 1930s. The employment

9 See Ernest R. Hilgard, ed., *American Psychology in Historical Perspective: Addresses of the Presidents of the American Psychological Association, 1892–1977* (Washington, D.C.: American Psychological Association, 1978), pp. 165–167, 399–400.

10 Data were derived from Hilgard, *American Psychology in Historical Perspective*. See also Robert S. Harper, "Tables of American Doctorates in Psychology," *American Journal of Psychology*, 1949, *62*:579–587.

picture, however, had changed dramatically, with an unmistakable shift toward public universities located in the Midwest. Employment in private institutions had declined from 90% in the prewar cohort to only 40% in the postwar group. And among members of the latter group, half were located in the Midwest, with the remainder split between the East (30%) and the West (20%). Occupational concentration was also more evident; two institutions, Illinois and Michigan, counted a total of seven presidents between them in the postwar group, after contributing none to the prewar group.[11]

The postwar leadership of the American Psychological Association remained largely in the hands of male psychologists, as before. No woman held the APA's highest office between 1921 (Margaret Washburn) and 1972 (Anne Anastasi). In 1951, after reviewing the association's record regarding women, Mildred Mitchell published a sharp critique of its evident sex discrimination. Even though women comprised more than one-third of its members, their representation in APA offices and on committees and boards was minimal.[12]

Mitchell's article in the *American Psychologist* raised the ire of Edwin Boring. Apparently still unsatisfied with the results of his wartime collaboration with Alice Bryan, he published "The Woman Problem" as a rejoinder to Mitchell.[13] It presented his views in their unadulterated form. Conveniently broadening the debate from one that affected only women to one that involved general social dynamics, Boring linked women's issues to the "great man in history" problem. By pointing out the universality of history's injustice to rank-and-file scientific workers, he may have intended to lessen the significance (and consequently the impact) of feminist complaints. In his inimitable style, Boring also offered an explanation for the continuing low status of women in the profession, emphasizing their lack of productivity and inability to work the legendary eighty-hour workweek that he considered the Harvard norm. The woman problem was thus a problem of "job-concentration." It existed because of the competition of "fanatics" (i.e., workaholics). This was not sex prejudice at all, he claimed, but realism. Boring's parting advice to women seeking higher prestige and status was (1) to write books, preferably with general, not applied, themes; and (2) to consider the effects of marriage. Although in his article Boring was careful only to imply that marriage might not be an asset to the professionally ambitious woman, he was devastatingly blunt in private: "If married, they may have more divided allegiance than the man. If unmarried, they have

---

11 Hilgard, *American Psychology in Historical Perspective.*

12 Mildred B. Mitchell, "Status of Women in the American Psychological Association," *American Psychologist,* 1951, 6:193–201.

13 Boring later recalled his work with Bryan: "The truth, however, was still left masked by compromises, until, some time after the collaboration was over, I published my own paper on the woman problem and felt that I had now said my own last word, even if not the very last one." Boring, *Psychologist at Large* (New York: Basic Books, 1961), p. 72.

conflict about being unmarried (although I did not say that. It seemed too infuriating to say.)"[14] For Boring, the woman problem hinged not on any identifiable discrimination but on mechanisms of acquiring prestige; although women faced some cultural obstacles, they were not insurmountable.[15]

More by default than by design, Boring may indeed have had the last word on the subject, at least for a time. His article marked the end of a period of public discussion over the role of women psychologists that had begun over a decade before with the outbreak of World War II. Between 1946 and the Mitchell–Boring exchange in 1951, only two articles explicitly concerned with women had appeared in the psychological literature. One was a postwar survey of their "work, training, and professional opportunities"; the other was a sarcastic account of one woman's job search.[16] Thus, by the early 1950s, after a decade of debate, the woman problem disappeared from the public discourse of psychology. Wartime mobilization had brought the issue of gender discrimination into focus, but fundamental changes proved impossible. Questions about the status of women were raised but not resolved, and they eventually faded from professional consciousness. Not until the late 1960s did a renewed feminist movement lead to significant reforms within the psychology community.[17]

Maintaining the Military Connection

In the dozen years following the war the number of psychologists employed by the military establishment increased greatly. In 1948 the APA listed 98 members (2% of the total membership) as working for various branches of the military services; by 1957 the number had increased to 729 (5%). Paradoxically, over the same period membership in the APA's Division of Military Psychology increased only slightly, from 153 to 249 individuals, which represented a proportional decline,

14 Boring to E. R. Hilgard, 10 August 1951; EGB/He-Hn 1951–52.
15 Edwin G. Boring, "The Woman Problem," *American Psychologist,* 1951, *6:*679–682.
16 H. A. Fjeld and L. B. Ames, "Women Psychologists: Their Work, Training, and Professional Opportunities," *Journal of Social Psychology,* 1950, *31:*61–94; Jane Loevinger, "Professional Ethics for Women Psychologists," *American Psychologist,* 1948, *3:* 551.
17 In the more supportive social context of the 1960s, with its growing emphasis on women's liberation, female psychologists founded the Association for Women Psychologists in 1969. Since then, it and associated groups have had a major impact in helping to eliminate the discrimination uncovered by women psychologists in the 1940s. See Virginia S. Sexton, "Women in American Psychology: An Overview," *Journal of International Understanding,* 1974, *9*(10):66–77; Elizabeth Scarborough, "Women in the American Psychological Association," in Rand B. Evans, Virginia Staudt Sexton, and Thomas C. Cadwallader, eds., *The American Psychological Association: A Historical Perspective* (Washington, D.C.: APA, 1992), pp. 303–325.

from 3% of the entire association to 1.6%.[18] Whatever the reasons for the disjunction between occupational status and professional affiliation, neither one provided an accurate gauge of the extent of military-psychology connections after the war.

The connection between psychology and the military carried over directly from the war. Nearly a quarter of the nation's 4,000 psychologists had been employed by the armed services during the war, and a substantial number of others had worked under military contracts. In a postwar survey, psychologists reported more satisfaction with their wartime utilization than any other scientific group.[19] Although most psychologists in the military demobilized and returned to civilian positions after the war, they proved quite willing to accept contracts and grants to work on problems of actual or potential relevance to the armed forces. For their part, defense agencies were eager to keep psychological expertise on tap and provided ample amounts of money for basic as well as applied research.

By 1948 the federal government was providing more than $2 million a year for extramural research in psychology in addition to its support of in-house efforts. Funds increased steadily to the point that in 1953 the total for all psychological research was over $10 million. For a decade after the war the Department of Defense was the largest single source of funds; in 1955, the Department of Health, Education, and Welfare took the lead (Figure 8.1).[20]

University psychology departments relied heavily on the Department of Defense to finance their expanding research programs. A 1951 survey revealed the full extent of this dependency. More than 40% of the entire research effort in psychology in American universities was being sponsored by defense agencies. (In comparison, nearly 70% of academic physics research was so supported.) Psychology ranked after physics, chemistry, electrical engineering, and aeronautical engineering in terms of the total manpower devoted to defense research.

Furthermore, at the leading institutions, defined here as those with ten or more

18 Calculated from 1948 and 1957 editions of the *American Psychological Association Directory.*
19 U.S. Army General Staff, *Scientists in Uniform, World War II* (Washington, D.C.: U.S. Army, 1948), pp. 52–53.
20 Marguerite L. Young and John T. Wilson, "Government Support of Psychological Research," *American Psychologist,* 1953, *8:*489–493; idem, "Government Support of Psychological Research: Fiscal Year 1954," *American Psychologist,* 1954, *9:*798–802; idem, "Governmental Support of Extramural Psychological Research: Fiscal Year 1955," *American Psychologist,* 1955, *10:*819–823; idem, "Government Support of Extramural Psychological Research: Fiscal Year 1956," *American Psychologist,* 1956, *11:*630–633; idem, "Government Support of Psychological Research: Fiscal Year 1957," *American Psychologist,* 1958, *13:*65–68; Marguerite L. Young and Henry S. Odbert, "Government Support of Psychological Research: Fiscal Year 1958," *American Psychologist,* 1959, *14:*497–500; idem, "Government Support of Psychological Research: Fiscal Year 1959," *American Psychologist,* 1960, *15:*661–664.

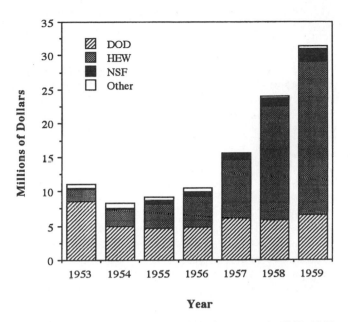

Figure 8.1. Federal support for psychological research, 1953–1959.

faculty-level researchers, fully half of the research effort was defense-related. Military patronage was not only a factor in the larger programs, however. Table 8.1, showing thirty-four universities, accounts for only about half of all military-funded research; the rest was distributed widely to dozens of other smaller academic programs around the country.[21]

One of the major sources of extramural funds during this period was the Office of Naval Research, which functioned as a "surrogate national science foundation" for several years following the war.[22] Through a generous system of contracts, it supported research in a number of areas, including psychometrics, training methods, group behavior, and human engineering. Probably three-quarters of the projects were based in universities, and the funds were used for graduate student support,

21 American Society for Engineering Education, Engineering College Research Council, Committee on Relations with Military Research Agencies, *University Research Potential: A Survey of the Resources for Scientific and Engineering Research in American Colleges and Universities* (Cambridge, Mass.: ECRC, June 1951). Data for the discipline of physics is presented in Paul Forman, "Behind Quantum Electronics: National Security as Basis for Physical Research in the United States, 1940–1960," *Historical Studies in the Physical and Biological Sciences,* 1987, *18:*149–229, which provided both the inspiration and the model for my analysis.

22 Harvey M. Sapolsky, *Science and the Navy: The History of the Office of Naval Research* (Princeton, N.J.: Princeton University Press, 1990), p. 118.

Table 8.1. *Leading academic performers of psychology research, 1951*

	Faculty-level researchers	FTEs	Defense-sponsored FTEs	% Defense	Grad student researchers	FTEs
Columbia University	17	10	10	100	30	17
City College of New York	11	2	2	100	23	5
Ohio State University	23	7.7	6.4	83	104	52
Northwestern University	15	6	5	83	16	5
Johns Hopkins University	15	9.5	7.5	79	14	5
University of California–Los Angeles	10	8	6	75	2	1
University of Rochester	14	7.5	5.5	73	25	10
University of Michigan	73	52	36	69	122	62
University of Washington	11	3	2	67	12	4
American University	13	13	7	54	20	20
Indiana University	15	5	2.5	50	40	15
Syracuse University	13	6	3	50	42	4
Brooklyn College	12	4	2	50	10	4
University of Florida	10	2	1	50	36	8
University of Illinois	24	14	6.75	48	24	14
Stanford University	14	5.5	2.5	45	35	9
Iowa State College	15	7.5	3	40	7.5	7.5
University of Southern California	15	6.5	2.5	38	70	45
Cornell University	16	8	3	37	35	12
Boston University	16.5	5.5	2	36	11	4
University of Minnesota	12	6	2	33	120	20
University of California	24	8	2	25	140	45
Purdue University	12	4	1	25	56	20
Duke University	11	4	1	25	32	15
Tulane University	10	4	1	25	27	7
Yale University	20	10	1	10	60	20
New York University	17	4	0	0	104	54
University of Nebraska	16	6	0	0	85	19
Michigan State University	14	4	0	0	29	7
University of Oregon	13	3.5	0	0	16	4.3
Temple University	13	2	0	0	10	4
University of Wisconsin	12	4	0	0	50	16
University of Buffalo	10	3	0	0	30	10
University of Kansas	10	2.5	0	0	75	15

Source: American Society for Engineering Education, Engineering College Research Council, *University Research Potential* (Cambridge, Mass.: ECRC, June 1951).

instrumentation, and facilities as well as normal research expenses. By 1956 the Office of Naval Research had disbursed nearly $18 million for psychological research since the war.[23]

In addition to the Office of Naval Research, the navy supported intramural research in the Psychology Section at the Naval Research Laboratory. The other military services also had their own in-house research groups as well as captive think tanks like the air force's RAND Corporation and the army's Human Resources Research Office. By 1953, a total of 690 psychologists, 40% with doctorates, were employed full-time by the military services.

RAND and Defense Science

In addition to funding a wide array of research on behavior, military interests also enlisted psychological expertise more directly. Along with other social scientists, psychologists became part of an expanding defense establishment after the war. The work of the RAND Corporation exemplified the new relationship forged between science and national defense as a result of the use of nuclear weapons. Strategic problems, once the province of military men steeped in military history, became subject to logical and empirical analysis by a new cadre of scientists and engineers. Operations research, systems analysis, and various other techniques and methods hybridized from mathematics, economics, psychology, engineering, and other fields became tools for the rational calculation of the costs and benefits associated with the development and deployment of new weapons. Complex scenarios were broken down into large sets of variables and subjected to mathematical modeling with the assistance of computers. New approaches such as cybernetics and game theory contributed to the developmenet of strategic nuclear studies at places like RAND and to the rise of an elite group of defense intellectuals dubbed the "wizards of Armageddon."[24]

RAND – an acronym for "Research and Development" – was begun as a cooperative project between the Douglas Aircraft Corporation and the air force. The head of the air force, General H. H. Arnold, had a keen appreciation of the importance of technological innovation in air warfare and wanted to continue the close working relationship the air force had developed with aircraft manufacturers during the war. "Project RAND" was organized within the Douglas firm in 1945 as a completely separate division that reported directly to the air force, not to company officials. The original contract called for "a program of study and research on the

23 John G. Darley, "Psychology and the Office of Naval Research: A Decade of Development," *American Psychologist,* 1957, *12:*305–323. For a review of ONR-supported research from 1945–1950, see Harold Guetzkow, ed., *Groups, Leadership and Men* (Pittsburgh, Pa.: Carnegie Press, 1951).

24 See Fred Kaplan, *The Wizards of Armageddon* (New York: Simon and Schuster, 1983).

broad subject of intercontinental warfare, other than surface, with the object of recommending to the army air forces preferred techniques and instrumentalities for this purpose."[25]

In addition to military officials, Project RAND had a set of civilian advisors, including San Francisco lawyer H. Rowan Gaither, who had served as the business manager of the Radiation Laboratory at MIT during World War II. Through his contact with Karl Compton, former director of the Radiation Laboratory and a trustee of the Ford Foundation, Gaither was asked to outline a program for the newly activated Ford philanthropy. As the report was being prepared, the Ford Foundation made a number of substantial grants, including one in 1948 for a million dollars to help transform Project RAND into the independent, nonprofit RAND Corporation.

While administrative arrangements were being worked out, RAND was hiring staff members "interested in some vague combination of mathematics, science, international affairs and national security" who enjoyed the intellectual atmosphere without the usual academic obligations of teaching and service.[26] In the fall of 1947, RAND was already home to about 150 civilian analysts when it sponsored a conference on social science. Held in New York City for the better part of a week, the meeting brought together thirty-five participants, including several leading social scientists and mathematicians.[27] Warren Weaver, director of the Rockefeller Foundation's Division of Natural Sciences and head of the wartime OSRD Applied Mathematics Panel, served as chairman.

In his opening remarks Weaver commented on the heterogeneity of the group:

> there's a curious assortment of individuals in this room at the present time –
> I didn't say an assortment of curious individuals, but a curious assortment
> of individuals. There is at least one person who is primarily an electrical en-
> gineer, one who is primarily an astronomer interested in meteors, two whose
> backgrounds are largely mathematical and one who is an aeronautical engi-
> neer. Having mentioned them, I think I have named most of those who come
> primarily from the physical sciences. Then there is the distinguished presi-
> dent of New Jersey Telephone Company, who is also a sociologist and many
> other things, and there are also a lot of you who represent, one after another,
> the disciplines, interests, and curiosities of the social sciences.[28]

Weaver went on to articulate the ideological ties that bound the group together. Foremost, he said, was a fundamental devotion to the rational life – "that there is

25 Paul Dickson, *Think Tanks* (New York: Atheneum, 1971), pp. 52–53.
26 Kaplan, *Wizards of Armageddon,* p. 62.
27 *Conference of Social Scientists, September 14 to 19, 1947 – New York* (Santa Monica, Calif.: RAND Corp. R-106, June 1948). Among the psychologists attending were Clark Hull (Yale) and Donald Marquis (Michigan).
28 Ibid., pp. 2–3.

something to this business of having some knowledge, and some experience, and some insight, and some analysis of problems, as compared with living in a state of ignorance, superstition, and drifting-into-whatever-may-come." Next came a shared concern over the state of the world, coupled with a desire to improve it: "In other words, I think that we are interested, not in war, but in peace; and that if there is anything that any one of us can do, by any possible effort he can bring to bear, he wants to do it." Weaver assumed that all the participants were "desperately dedicated to the ideals of democracy" and would do anything for national defense in case of a "showdown."[29]

After Weaver's passionate Cold War rhetoric, the conference dealt with five broad areas of social science: psychology and sociology; political science; economics; intelligence and military affairs; and research methods, organization, and planning. After general topics were outlined for possible consideration by RAND, specific proposals were made for research projects. An example of the kinds of psychology projects supported by RAND was a study of the psychological aspects of civil defense conducted by Yale psychologist Irving Janis, who received a contract in 1949. His report, a brief version of which was published in the *Bulletin of the Atomic Scientists,* contained recommendations on how to minimize the negative emotions and behaviors that would occur in the wake of an atomic attack.[30]

Through its in-house research and extramural funding, RAND developed into a major defense think tank during the Cold War. Psychologists became part of such efforts, and, like their colleagues in other social science disciplines, found such work congenial. It was an exciting continuation of the interdisciplinary, mission-oriented research from the war, amply funded and intellectually rewarding. It provided yet another employment venue for psychologists uninterested in or unsuited for traditional academic teaching. For leaders in the profession, it provided access to high-level governmental and scientific circles.

The Rise of Clinical Psychology

One of the cornerstones of psychology's postwar expansion was laid at the end of the war when the Veterans Administration and the Public Health Service began programs to support research and graduate training in clinical psychology. Responding to shortages of trained personnel to minister to the needs of thousands of veterans with "neuropsychiatric" problems, the Veterans Administration practically created a new mental health specialty by providing training funds as well as

---

29 Ibid., p. 3.
30 Paul Boyer, *By the Bomb's Early Light: American Thought and Culture at the Dawn of the Atomic Age* (New York: Pantheon, 1985), pp. 331–333, which cites Irving L. Janis, "Psychological Problems of A-Bomb Defense," *Bulletin of the Atomic Scientists,* 1950 (Aug.–Sept.), pp. 257–259.

hospital internships and permanent jobs for clinical psychologists. The massive infusion of funds for research and training transformed university psychology departments, causing a rapid rise in the number of faculty members, students, and support staff as well as major changes in curriculum.

The war had created the conditions necessary for the full institutionalization of clinical psychology. Before then, despite several attempts to create a graduate-level clinical track in the standard psychology curriculum, clinical psychologists, like other practitioners, developed most of their skills through informal internships and on-the-job training. As late as 1940 it was impossible to obtain a formal Ph.D. in clinical psychology, even though there were dozens of psychologists working as clinicians.

During the war a strong demand for clinical psychologists developed, and it became clear that the supply of trained personnel was lacking. The army had sought to employ nearly 350 clinical psychologists but had been able to obtain the services of only 250 by the end of the war.[31] In September 1945 the Office of the Chief Clinical Psychologist was transferred from the Adjutant General's Office to the Surgeon General's Office, and clinical psychologists became part of the Medical Administrative Corps, thus consolidating administrative and professional control of the program under medical auspices.[32] Military demobilization proceeded quickly; by April 1946 fewer than 70 clinical psychologists remained on active duty.[33] Plans were made to organize an Association of Military Clinical Psychologists and to publish a newsletter in order to encourage professional communication.[34] The Advisory Board on Clinical Psychology held its last meeting and made its final recommendations, urging the continued utilization of clinical psychologists and the creation of a strong research program. A research program was necessary, Robert Sears argued, in order to attract high-quality personnel and to keep clinicians from getting stale.[35]

At the end of the war military casualties began to fill the hospitals of the Veterans Administration. Among those needing treatment were more than forty thou-

---

31 The procurement of 346 clinical psychologists had been authorized; 305 for the Service Forces, 35 for the Air Forces, and 6 for the Office of Strategic Services. A. J. Glass and R. J. Bernucci, eds., *Neuropsychiatry in World War II,* vol. 1 (Washington, D.C.: Surgeon General's Office, Department of the Army, 1966), pp. 573–574.

32 Ibid., p. 37.

33 One survey counted 50, while another source noted 65; Max Hutt and Emmette O. Milton, "An Analysis of Duties Performed by Clinical Psychologists in the Army," *American Psychologist,* 1947, 2:52–56; Robert R. Sears (Notes on ABCP meeting), 19 April 1946; RRS/4.

34 Max L. Hutt to Sears, 14 March 1946; RRS/4.

35 ABCP, "Minutes of Third Meeting," 19 April 1946; WVB/24/CCMP, ABCP. Sears was also concerned that practical demands might overwhelm research; Sears to Bingham, 25 April 1946; RRS/4.

sand neuropsychiatric patients. In order to cope with the large caseload, the VA looked to psychologists. In late 1945, George A. Kelly, a former navy clinical psychologist who had joined the University of Maryland faculty, became a consultant. The VA established the Clinical Psychology Program in the Bureau of Medicine and Surgery in 1946. It was under the direction of James Grier Miller, a former OSS officer who held both M.D. and Ph.D. degrees from Harvard. The most significant part of the VA program was the establishment of a Clinical Psychology Training Program, which provided funding for the graduate education of students specializing in clinical psychology.[36]

Begun in the academic year 1946–1947, the program involved 22 universities at the start and funded 200 trainees. The program proved to be a boon to some graduate psychology departments and encouraged the development of training standards and accreditation procedures by the American Psychological Association. After three years of operation, VA support extended to more than 1,500 students in 50 institutions around the country.[37]

The VA program evolved along with professional training standards and accreditation procedures developed by the American Psychological Association. One of the first official acts of the newly reconstituted APA was to form a Committee on Graduate Training in 1946. This group was to formulate standards for the accreditation of university programs, and to provide information and advice to the Veterans Administration. Through such advisory bodies, the federal government relied on the psychology community to help organize and administer funding programs. In turn, the psychology community depended upon government support to assist in defining its role as a mental health profession.

In 1949 the Public Health Service provided a major grant to the APA to underwrite a Conference on Clinical Training in Psychology. Dubbed the "Boulder Conference" because of its location at the University of Colorado, the meeting gathered 71 participants together for two weeks to review the rapid growth of clinical training programs and to chart future developments. In its report, the Boulder Conference reaffirmed the central role in graduate education played by university psychology departments, and emphasized the advantages in training students for both

---

36 James G. Miller, "Clinical Psychology in the Veterans Administration," *American Psychologist,* 1946, *1:*181–189.
37 Harold M. Hildreth, "Clinical Psychology in the Veterans Administration," in E. A. Rubenstein and M. Lorr, eds., *Survey of Clinical Practice in Psychology* (New York: International Universities Press, 1954), pp. 83–108. For an overview of later developments, see Dana L. Moore, "The Veterans Administration and the Training Program in Psychology," in Donald L. Freedheim, ed., *History of Psychotherapy: A Century of Change* (Washington, D.C.: American Psychological Association, 1992), pp. 776–800; Rodney R. Baker, "VA Psychology, 1946–1992" (Association of VA Chief Psychologists, 1992), 4 pp.

research and practice. Although the "scientist-practitioner model" was not fully articulated at the conference, it soon emerged as the dominant conception of the proper role of the clinical psychologist.[38] It could hardly have been otherwise, given the disciplinary investment in the scientific status of psychology and the institutional infrastructure built up around laboratory research.

Although its implications were not fully appreciated at the time, the decision to graft clinical training onto the existing graduate curriculum in experimental psychology proved fateful. Among its consequences was a rapid response to new sources of federal training monies provided by the Veterans Administration and the Public Health Service. Psychology departments lost little time in reorganizing their graduate programs to comply with the requirements of such initiatives. This approach also prevented a significant increase in the overall completion time for the doctorate, providing a significant advantage over psychiatry, with its postgraduate residency training. Furthermore, the incorporation of clinical programs into academic departments provided an important mechanism for harmonizing the indoctrination of aspiring psychologists, whatever their occupational destination. The ethos of experimentalism continued as the foundation of professional ideology for future researchers and therapists alike. Finally, as psychology departments grew in size, their influence within the university increased proportionately.

By the mid-1950s the effects of federal support for clinical psychology were quite apparent. In 1954 there were about 1,500 members of the American Psychological Association's Division of Clinical Psychology, nearly twice as many as the combined membership of its Divisions of Experimental Psychology and General Psychology. In the same year, the Veterans Administration employed 476 staff psychologists, all with doctorates. More than half of them had come via the VA training program established only eight years before.[39] The growth of clinical psychology continued and by the early 1960s had reshaped the structure of its parent discipline to the point that a majority of psychologists were employed in nonacademic settings, reversing a tradition dating back to the turn of the century.[40]

The Evolving Support System

As psychologists reconfigured their professional identity and modes of voluntary association, their extrinsic support system underwent massive changes as well. Overall federal expenditures for science skyrocketed during the war and contin-

---

38 Victor C. Raimy, ed., *Training in Clinical Psychology* (New York: Prentice-Hall, 1950), pp. 195–196.

39 Three out of five were alumni of the VA training program; Hildreth, "Clinical Psychology in the Veterans Administration," pp. 86, 104.

40 Robert C. Tryon, "Psychology in Flux: The Academic–Professional Bipolarity," *American Psychologist,* 1963, *18:*134–143.

ued to grow well into the postwar era. Like most disciplines, psychology benefited from this unprecedented flood of dollars for scientific research, application, and training. Some fields, such as physics, received support on the basis of their links to a highly specific national priority – in this case, national security. Indeed, support for the physical sciences more generally was justified on a highly elastic notion of their relevance to national defense. The situation in the life sciences was different. The rationale for federal involvement in this sector drew on the perception that basic research in biology and medicine would contribute to the improvement of national health. Funding for social science presented yet another pattern of support as the government continued to rely on its expertise to manage various programs related to human welfare.

Shortly before the end of the war the appearance of Vannevar Bush's report, *Science – The Endless Frontier,* focused attention on how the government might continue to support science in peacetime.[41] Bush, as head of the Office of Scientific Research and Development, was anxious to make a smooth transition from the temporary wartime organization of support to more permanent arrangements. For him, a key issue was how to balance the intellectual freedom and autonomy deemed necessary for scientific investigation with the government's understandable interest in accounting for the expenditure of tax dollars.

The lesson of the war was clear: "The Federal Government should accept new responsibilities for promoting the creation of new scientific knowledge and the development of scientific talent."[42] A major premise of the Bush report was that basic or pure research was the source of technological innovation: "Basic research leads to new knowledge. It provides scientific capital. It creates the fund from which the practical applications of knowledge must be drawn. New products and new processes do not appear full-grown. They are founded on new principles and new conceptions, which in turn are painstakingly developed by research in the purest realms of science."[43] It followed, then, that it was necessary to replenish the reservoir of knowledge and talent systematically in order to continue to enjoy the fruits of research.

Bush proposed the creation of a National Research Foundation as the "mechanism" whereby the government would fulfill its enlarged responsibilities toward science. Dedicated to the support of basic science, this new agency would provide extramural funds, mainly to colleges and universities and nonprofit research institutes, to pursue research in the physical sciences, medicine, and military matters, and to develop a program of fellowships and scholarships. By focusing exclusively on basic rather than applied research, the new agency would complement existing governmental laboratories and programs concerned with science in relation to operational problems, as in agriculture or commerce.[44] Clearly aimed at the physical

41 Vannevar Bush, *Science – The Endless Frontier* (Washington, D.C.: GPO, 1945).
42 Ibid., p. 31.   43 Ibid., p. 19.   44 Ibid., pp. 34–40.

and biomedical sciences, *Science – The Endless Frontier* omitted discussion of the social sciences.

Bush's report was widely circulated and became a touchstone for the congressional debates that had already begun over postwar science policy.[45] It articulated a vision of government support that allowed the scientific community a large measure of autonomy. Elitist and politically insulated, it was a "best-science" approach to allocating funds. An alternate perspective was provided by Senator Harvey Kilgore, a New Dealer from West Virginia, who had become interested in science policy issues during the war and proposed the creation of a National Science Foundation following the war. In addition to supporting basic research and advanced training, Kilgore's scheme called for the coordination of all federal research activities with an eye toward achieving social benefits. Support for the social sciences was included, and at least part of the funds were to be distributed on a geographical basis, thus spreading federal largesse to all parts of the country.

Conservatives, including Bush, reacted negatively to Kilgore's liberal proposal. They objected to the inclusion of the social sciences, which the more vociferous among them regarded "as just so much political propaganda masquerading as science."[46] In 1946 a bill sponsored by Kilgore passed the Senate, after its clause calling for support of social science was dropped, while a Bush-inspired measure was considered by the House of Representatives. Unable to reconcile the two bills, Congress adjourned without creating a national science foundation.

Over the next several years congressional debate continued without closure until 1950, when an act establishing the National Science Foundation was finally passed. As created, the foundation represented a victory for Bush's position, although it did contain a permissive clause that would allow for the inclusion of the social sciences. In the meantime, however, an ad hoc system of research support had emerged, dominated by the military. In this context, the initial appropriation of $350,000 for the National Science Foundation, which was limited to an annual maximum of $15 million, was dwarfed by the expenditures of the Public Health Service, the Atomic Energy Commission, and the Department of Defense, which together totaled $63 million in 1949.[47]

Debate Over Social Science in the NSF

Organized lobbying for the inclusion of the social sciences in postwar federal funding programs came relatively late in the process and proved ineffective. The ab-

---

45 Daniel J. Kevles, "The National Science Foundation and the Debate over Postwar Research Policy, 1942–1945: A Political Interpretation of *Science – The Endless Frontier*," *Isis,* 1977, *68*:5–26. See also Kevles, *The Physicists* (New York: Knopf, 1978), pp. 343–348, 356–366.

46 Kevles, *The Physicists,* p. 345.   47 Ibid., p. 358.

sence of the social sciences at the start of the National Science Foundation in 1950 is less a story of failure, however, than a tale of alternative funding opportunities, especially in the field of psychology.[48] This episode points up the differences in patterns of patronage between the natural and the social sciences that had evolved over a long period, and that were in some but not all key respects altered by wartime mobilization. Psychology, strategically located at the nexus of natural, biomedical, and social science, straddled the important organizational divide represented by the National Research Council and the Social Science Research Council. Its major professional organization, the American Psychological Association, maintained membership in both groups.

Begun in 1923, the Social Science Research Council (SSRC) was a voluntary association composed of the disciplinary societies of anthropology, economics, history, political science, psychology, sociology, and statistics. In February 1945 it appointed a Committee on the Federal Government and Research, which included nationally prominent scientists such as Charles Merriam, Wesley Mitchell, and Robert Yerkes. This group, composed of senior statesmen of social science, was fairly content with the status quo, having already successfully engineered viable professional relationships with the federal government, on an individual as well as collective basis. Its original recommendation, issued before the Bush report, echoed conventional mistrust of government support for academic research expressed by the National Resources Planning Board. Within a few months, however, the appearance of *Science – The Endless Frontier* caused the committee to revise its position and joined the growing bandwagon for the inclusion of the social sciences in a national science foundation.[49]

Commissioned by the SSRC to draft a report, the committee responded quickly and outlined several key issues facing the social sciences, including the possible influence of government funding on the scope and direction of research, as well as how federal support might best be administered. As Congress considered legislation for a national science foundation in 1946, the SSRC and other groups of social scientists became more active. The American Sociological Society appointed a committee to study the issue. It was chaired by Harvard sociologist Talcott Parsons, who drafted a position paper that circulated widely, first in an unpublished form and then in various printed versions.[50]

48 A recent dissertation covers government and private foundation support for social science after World War II but neglects psychology. It is an excellent source for the wider context of debates over the role of social science in the U.S.A. See Mark Solovey, "The Politics of Intellectual Identity and American Social Science, 1945–1970" (Ph.D. diss., University of Wisconsin, 1996).

49 Samuel Z. Klausner, "The Bid to Nationalize American Social Science," in Samuel Z. Klausner and Victor M. Lidz, eds., *The Nationalization of the Social Sciences* (Philadelphia: University of Pennsylvania Press, 1986), pp. 3–39, on p. 5.

50 Talcott Parsons, "The Science Legislation and the Role of the Social Sciences," *American*

In his analysis of the failure of legislative action during the 79th Congress, Parsons suggested that disagreement over the inclusion of the social sciences was not a primary factor. Rather, the key issues revolved around administrative structure and patent policies.[51] In his attempt to construct a case for federal support of social science research, Parsons stressed three overarching themes. First, the need to explore how science might contribute to the solution of social problems. Second, the importance of developing social science as a further expression of the values of rational inquiry and understanding. Third, a belief in the fundamental unity of all science, both natural and social, impels a comprehensive and coordinated effort across different fields. He went on to cite the need for more empirical work and the gathering of social data, and for close interactions between such activities and the construction of theory. Parsons argued that the success of applied social research in World War II demonstrated that, given the appropriate human and material resources, "the social sciences can deliver . . . results which are of direct and fundamental practical importance."[52] He concluded, however, that the case for support rested more on the "potentialities" of the social sciences than on their past achievements.[53]

As debate over the proposed science foundation continued, in late 1946 the SSRC recast its Committee on the Federal Government and Research, appointing Robert Yerkes as chairman and several new members, including Talcott Parsons and Dael Wolfle, recently installed as the executive secretary of the American Psychological Association. The group was charged with the task of preparing a "white paper" on the social sciences that would be analogous to Bush's blueprint for the natural sciences, and the job of drafting the report fell to Parsons.

Pressed with other responsibilities, including the establishment of the Department of Social Relations at Harvard, Parsons was unable to complete even a rough draft until two years later. Entitled *Social Science: A Basic National Resource,* the manuscript was reviewed by SSRC officials, who found it wanting. It was sent to dozens of reviewers, whose reactions ranged from strong criticism to muted disappointment. Highly academic in tone, it was filled with the complex conceptualizations and characteristic jargon of Parsons's other writings. Some thought the thrust was too elitist and too focused on the professionalization of social science and the development of technical knowledge. Others criticized its neglect of the "political-economic-historic complex" in favor of sociology, psychology, and anthropology – precisely those disciplines that formed the basis of the Harvard De-

Sociological Review, 1946, *11*:653–666; idem, "National Science Legislation, Part 1: An Historical Review," *Bulletin of Atomic Scientists,* 1946, *2*(10):7–9; National Science Legislation, Part 2: The Case for the Social Sciences," idem, 1947, *3*(1):3–5.

51 Parsons, "National Science Legislation, Part 1," p. 8.
52 Parsons, "National Science Legislation, Part 2," p. 5.
53 "Parsons perceived his role as an architect of an interdisciplinary empirical social science." Klausner, "Bid to Nationalize," p. 14.

partment of Social Relations (and that in other circles were beginning to be referred to as the "behavioral sciences"). It was clear that something radical had to be done in order to salvage the report.[54]

Parsons was asked to revise the document, but other activities took priority. Another year passed, and in September 1949 the SSRC decided that the report might be rescued by calling in a collaborator. John W. Riley, chairman of the Rutgers sociology department, was an obvious candidate. He was already working on a study for the Ford Foundation, then in the process of developing its overall program, on "what we know about the social sciences." In contrast to Parsons, Riley concentrated on the organization, funding, and training of social scientists. The collaboration led to a joint draft manuscript entitled *The Status of the Social Sciences*, delivered to the SSRC in 1951, a year following the establishment of the National Science Foundation, which had not included the social sciences among the fields it supported.[55]

Although the social sciences as a group were not included in the National Science Foundation, some psychological and anthropological research areas close to biology did receive funding. In fact, the fate of psychology in the proposed foundation was more closely tied to provisions for the natural sciences than the social sciences. Emphasizing its biological connections, psychology found a place among the initial disciplines supported. It was indicative that APA Executive Secretary Dael Wolfle served as secretary of the Inter-Society Committee for a National Science Foundation, an effort by the scientific community to present a united front, organized by the AAAS.[56]

Support for psychology from the National Science Foundation had ideological as well as financial ramifications. It demonstrated that psychology, or at least segments of it, belonged in the circle of the natural sciences. Through his work on the Intersociety Committee, Wolfle not only displayed his own administrative talent but also exemplified the entrance of psychologists on the stage of federal science policy-making.[57]

---

54 Ibid., pp. 24–27.
55 Ibid., pp. 27–30. An edited draft of the report Parsons submitted to the SSRC in 1948 was published (for the first time) as Talcott Parsons, "Social Science: A Basic National Resource," in Klausner and Lidz, eds., *The Nationalization of the Social Sciences,* pp. 41–112.
56 Dael Wolfle, "Making a Case for the Social Sciences," in Klausner and Lidz, eds., *The Nationalization of the Social Sciences,* pp. 185–196. Dael Wolfle, "The Inter-Society Committee for a National Science Foundation," *Science,* 1947, *106:*529–533. See Gene M. Lyons, *The Uneasy Partnership: Social Science and the Federal Government in the Twentieth Century* (New York: Russell Sage Foundation, 1969), pp. 126–136, which notes that the "hard edge" of social science, experimental psychology, was a program area within the biological sciences at the time social science programs were being organized after 1953 (p. 271).
57 After leaving the APA in 1950, Wolfle directed the federal Commission on Human

Private Foundation Support

The war caused a dramatic increase in federal funds for research, especially in the physical and biomedical sciences. Although the social sciences shared in this largesse, the lion's share went to fields more closely associated with national defense and health services. Before the war, in 1938, about a quarter of federal research dollars went to the social sciences. By 1952 that portion had dwindled to 2%, despite the fact that federal social science budgets tripled between 1938 and 1952, from $17 million to $55 million. Surging budgets for the natural sciences, however, transformed this absolute increase into a relative decline.[58]

As the federal government emerged as the prime patron of the natural sciences, its support eclipsed that provided by private foundations, previously an important source of funds. In contrast, private foundation patronage remained important to the social sciences, including psychology, after the war.[59]

Less expensive and less massive than the physical sciences, the social sciences garnered significant federal resources while at the same time remaining a favorite investment for private foundations. In fact, foundations were probably the most important source of funds for basic research in the social sciences in the decade after the war. Between 1945 and 1954 the "big three" – the Rockefeller Foundation, the Carnegie Corporation, and the Ford Foundation – contributed more than $40 million for social science research. Motivated by concerns over society and ethics in the wake of the atomic bomb, foundation officials found ready justification for supporting such research. Many perceived a growing gap between the ability of humans to unleash the forces of nature and their skill in making them serve humanitarian ends. And foundation interest in social science was no doubt reinforced by the presence of many trained social scientists in the world of philanthropy.

At the Rockefeller Foundation, the end of the Second World War brought a new urgency to President Raymond Fosdick's perennial plea for social reform. The scientific developments of wartime, which had contributed new horrors to modern

Resources and Advanced Training for four years until beginning his long tenure (1954– 1970) as Executive Officer of the American Association for the Advancement of Science.

58 Milton D. Graham, *Federal Utilization of Social Science Research: Exploration of the Problems* (Washington, D.C.: Brookings Institution, 1952), p. 48, table 5. A later estimate revised the figures to 92% for natural science and 8% for social science in 1952. Kathleen Archibald, "Federal Interest and Investment in Social Science," in *Social Research in Federal Domestic Programs* (Washington, D.C.: GPO, 1967), pp. 314–341, on p. 329.

59 See Robert E. Kohler, *Partners in Science: Foundations and Natural Scientists, 1900– 1945* (Chicago: University of Chicago Press, 1991); Roger L. Geiger, "American Foundations and Academic Social Science, 1945–1960," *Minerva,* 1987, *25:*315–341.

warfare, provided fresh imagery for the notion of cultural lag. Fosdick declared, in his 1945 review of foundation activities, that

> Men are discovering the right things but in the wrong order, which is another way of saying that we are learning how to control nature before we have learned how to control ourselves. . . . There is no penicillin, no sulfa drug, for the sickness which afflicts our civilization. No social or ethical atomic bomb can be devised to neutralize the weapons with which we have armed our savage instincts.[60]

After the war, the Rockefeller Foundation's social science program began to concentrate on international relations studies and on academic research and training. Among the recipients of large grants in 1945 were the new Russian Institute at Columbia ($250,000), the Royal Institute of International Affairs in London ($196,000), and the Foreign Policy Association of New York ($200,000). The Social Science Research Council was provided $300,000 for fellowships in the social sciences.

In 1946 Fosdick reiterated his appeal for more social research:

> While we cannot put brakes on intellectual adventure, it must be admitted that there is a lack of balance about our studies and our research that imperils the future. The disproportion between the physical power at our disposal and our capacity to make good use of it is growing with every day that passes. We are in the midst of a revolution in our physical environment so vast and so rapid that our minds can scarcely keep up with it. But there are other things that cannot keep up with it, either – notably our social ideas, our habits of life and our political and economic institutions. . . . Our knowledge of human behavior and social relations is not adequate to give us the guidance we need; and the fundamental issue of our time is whether we can develop understanding and wisdom reliable enough to serve as a chart in working out the problems of human relations; or whether we shall allow our present lopsided progress to develop to a point that capsizes our civilization in a catastrophe of immeasurable proportions.

To Fosdick, it was clear that specialized and coordinated expertise was needed to avoid the destructive capabilities unleashed by the atomic bomb. Research would provide the basis for new knowledge about human relations, knowledge that was necessary in order to understand and master "social nature" in a manner analogous to the mastery of physical nature that had been demonstrated in the war.[61]

In spite of Fosdick's inspired rhetoric, and a mandate from the Trustee's Committee in 1946 to promote basic research in social anthropology, social psychology,

---

60 Raymond Fosdick, "President's Review," *Rockefeller Foundation Annual Report,* 1945.
61 Raymond Fosdick, "President's Review," *Rockefeller Foundation Annual Report,* 1946.

and sociology, the Rockefeller Foundation did not venture into new territory in its postwar support of social science. Nor did it increase the portion of its annual program funds going to the social sciences, which remained at less than 20% of the total during the postwar decade. It continued its traditional program of large institutional grants and support for individuals in the form of research grants and fellowships. Between 1945 and 1955 the major beneficiaries of the foundation's support in the social sciences were three institutions that it had supported for a long time: the Social Science Research Council, the National Bureau of Economic Research, and the Brookings Institution. Individual research grants and postdoctoral fellowships accounted for another substantial portion; the remainder was divided among various grants made to various academic institutes.[62]

Like the Rockefeller Foundation, the Carnegie Corporation had long-standing interests in the social sciences. Possessing smaller assets, it averaged about $6 million in annual appropriations between 1945 and 1955. Under the generally conservative leadership of Frederick Keppel (1923–1942), the foundation commissioned Gunnar Myrdal's pathbreaking research on African Americans and provided $300,000 in funds beginning in 1938. That landmark study, published as *The American Dilemma* in 1944, demonstrated how empirical social research could contribute to social reform efforts.[63]

In his report for 1946, President Josephs noted that the war had demonstrated the importance of social science, as well as the need to train additional personnel. He observed a "lack of balance" in education toward the natural sciences and suggested that the social sciences and the humanities needed promoting. Policy shifted accordingly, and by 1948 the proportion of foundation funds appropriated for the social sciences had increased substantially. That year the foundation made two of its largest grants: $750,000 to help set up the Educational Testing Service and $740,000 to establish the Russian Research Center at Harvard.[64]

Two small grants made in 1947 displayed the foundation's role in providing seed

---

62 Between 1945 and 1955 the Rockefeller Foundation expended a total of $183.8 million, including a total of $32.5 million donated by the social science division. The SSRC received more than $2 million, the National Bureau of Economic Research over $4 million, and the Brookings Institution $1.3 million. Grants to individuals totaled $3.29 million; fellowship funds $3.5 million. Other notable grants were made to the University of Chicago Division of Social Sciences ($150,000 in 1945), Harvard Laboratory of Social Relations ($100,000 in 1949 and again in 1951), and a Yale project on communication and attitude change ($200,000 in 1954). Calculated from *Rockefeller Foundation Annual Report,* 1945–1955.

63 See Walter A. Jackson, *Gunnar Myrdal and America's Conscience: Social Engineering and Racial Liberalism, 1938–1987* (Chapel Hill: University of North Carolina Press, 1990).

64 Carnegie Corporation Annual Report, 1946 and 1948.

money. Stuart Chase, a popular writer on economics and public affairs, received $7,500 to produce a survey of the prospects of the social sciences in the wake of World War II. Published as *The Proper Study of Mankind* (1948), the book asserted that "the startling success of the Manhattan project has forced social scientists to come of age." Sounding again the familiar theme that existing patterns of human relationships had been made obsolete by technological advances, Chase probed the findings and methods of social science research in hope of solutions to the problems to human conflict. He suggested greater financial and moral support for the central fields of anthropology, psychology, sociology, economics, and political science, as well as renewed efforts to integrate and correlate their findings. He was cautiously optimistic that if social scientists emulated their counterparts in the natural sciences, ways could be found to help humans come to terms with each other as well as with nature.[65]

In 1947 the Carnegie Corporation gave a $10,000 grant to Donald Marquis, chair of the psychology department at the University of Michigan, for consultation with leading social scientists in order to produce a "fresh appraisal of the place and the functions of the social sciences."[66] Already well along in his efforts to build an academic empire in psychology and allied fields at Michigan, Marquis had emerged as an influential figure in national circles concerned with social science policy and research funding. During the presidency of sociologist Charles Dollard between 1948 and 1954, the philanthropy continued its social science program. But its new leader sounded a cautionary note about the "brave verbalizations" of grand theory in the social sciences and argued that a more focused interdisciplinary attack on specific problems would be more fruitful, such as the work on consumer behavior undertaken by Michigan's Survey Research Center. After psychologist John Gardner (a veteran of the OSS assessment program in World War II) became president in 1954, he reoriented the foundation's program around the problems of American education, particularly higher education.[67]

With the death of Henry Ford in 1947, the quiescent philanthropic foundation he had established several years earlier was roused into robust life. His huge fortune flowed into the coffers of the Ford Foundation, making it the largest private foundation in American history. In formulating plans to disburse its funds, foundation trustees published the *Report of the Study for the Ford Foundation on Policy and Program* in 1949. Prepared under the direction of San Francisco lawyer H. Rowan Gaither, the wartime business manager of MIT's Radiation Laboratory, it outlined

---

65 Stuart Chase, *The Proper Study of Mankind . . . An Inquiry into the Science of Human Relations* (New York: Harper, 1948), quoted on p. 5.
66 Carnegie Corporation Annual Report, 1947, p. 32.
67 See Ellen Condliffe Lagemann, *The Politics of Knowledge: The Carnegie Corporation, Philanthropy, and Public Policy* (Chicago: University of Chicago Press, 1989).

five main areas of concentration. Four directly addressed issues of human welfare connected to peace, democracy, the economy, and education. The fifth entailed a program to "support scientific activities designed to increase knowledge of factors which influence or determine human conduct, and to extend such knowledge for the maximum benefit of individuals and of society."[68] As large grants were being made in the first four program areas, the development of Program Area V, "Individual Behavior and Human Relations," was evolving in consultation with leading academic social scientists, including Donald Marquis, who served as a key consultant.

As the Ford Foundation faced the challenges of giving away unprecedented funds, social scientists were busy trying to convince them to channel their money into existing centers of activity. The foundation sponsored self-studies at five leading universities to help determine opportunities and needs in the area. As plans evolved, the term "behavioral science," first used at the University of Chicago by James Grier Miller and his colleagues to describe their interdisciplinary approach, became attached to the program. With psychology, anthropology, and sociology at the core, behavioral science suggested the goal of an integrated human science, unified by a common focus on "behavior," at whatever level of individual or collective action.

After several years of planning and preliminary grant making, the Ford Foundation decided to establish an independent think tank. In 1952 the creation of the Center for Advanced Study in the Behavioral Sciences was announced, with an initial endowment of $3.5 million. Eventually located near Stanford University, the center was envisioned as an "academic oasis" where a few dozen scholars could come and spend a fellowship year pursuing their work without the usual distractions and in the company of stimulating colleagues. By creating a new center with a rather elastic definition of its bailiwick, the foundation avoided direct competition with universities and among disciplines. Good people, not particular fields, were to be supported. In opting for this organizational model, the Ford Foundation sidestepped the issue of research training and apprenticeship that had been raised as a key issue in the improvement of the behavioral sciences. It also harked back to a more scholarly image of the researcher, as no laboratories or similar facilities were provided.[69]

Endorsing the notion of the behavioral sciences enabled the Ford Foundation simultaneously to avoid some of the pejorative connotations of "social" science

68 Study Committee, *Report of the Study for the Ford Foundation on Policy and Program* (Detroit, Mich.: Ford Foundation, November 1949), quoted on p. 90.
69 Arnold Thackray, "CASBS: Notes Toward a History," *Center for Advanced Study in the Behavioral Sciences Annual Report 1984,* 1984, pp. 59–71; idem, "A Site for CASBS: East or West?" *Center for Advanced Study in the Behavioral Sciences Annual Report 1987,* 1987, pp. 63–71.

and to embrace the latest trends in interdisciplinary human science. Conceptually as well as rhetorically, behavioral science reaffirmed the centrality of psychology to any comprehensive science of human nature. Thus it is not surprising that the field and its practitioners were among the main beneficiaries of the foundation's largesse.

The Center for Advanced Study in the Behavioral Sciences, like the Institute of Human Relations at Yale three decades earlier, embodied the hopes and dreams of academic social scientists. Both attracted generous support, which allowed debates about their purpose and shape to take place relatively unconstrained by many of the quotidian pressures usually encountered in academic life.[70]

Additional grants in the behavioral sciences were made in 1955 as an extension of the earlier program of university surveys. Harvard, Stanford, North Carolina, Michigan, and Minnesota each received more than $100,000 for research and training support. Harvard's Department of Social Relations, for instance, got $400,000 for faculty development in the field of personality psychology; part of Michigan's $220,000 total went to the Institute for Social Research for large-scale survey; and Stanford's $400,000 was slated mainly for hiring faculty specializing in communications, mathematics, and social psychology. The foundation also disbursed up to a hundred grants-in-aid to individual researchers. All told, nearly $2 million went to support the behavioral sciences in 1955.[71]

Conclusion

Academic psychology prospered with plentiful federal support as university departments produced more research and more researchers. Various military groups supported academic research in areas with potential military relevance, which was often defined quite broadly. In addition to helping the government figure out how to fight the next war, psychologists were involved in rehabilitating casualties from the last one. The high incidence of neuropsychiatric problems among World War II veterans created a demand for psychological diagnostic and therapeutic services that was filled by newly trained clinical psychologists, whose education was largely underwritten by the Veterans Administration and the U.S. Public Health Service. As the federal government expanded its programs in defense, health, education, and social welfare, psychology's protean identity enabled its practitioners to find allies practically everywhere and to garner support from a variety of sources.

70 See James H. Capshew, "The Yale Connection in American Psychology: Philanthropy, War, and the Emergence of an Academic Elite, 1930–1955," in *The Development of the Social Sciences in the United States and Canada: The Role of Philanthropy,* eds. Theresa R. Richardson and Donald Fisher (Greenwich, Conn.: Ablex Publishing Corporation, 1998).

71 *Ford Foundation Annual Report, 1955,* pp. 35–37.

The university psychology department remained the basic institutional unit of the discipline after the war, continuing to play a central role in the production of new scientists as well as new science. Its flexible structure made it easy for most departments to graft clinical research and training onto their existing programs. Grants, contracts, and other forms of extramural support were also folded into institutional budgets, with few problems. But local conditions, disciplinary priorities, and outside support combined in various ways meant that departmental organization took many forms in response to particular pressures. The postwar evolution of several major university departments shows how the influx of new resources – cognitive, technical, and financial – affected the identity and image of the field as it was refracted through the prism of local academic culture.

In keeping with the ideological and organizational unity prompted by the war, efforts to refurbish psychology's academic home took for granted the advantages of solidarity and cohesion. Growth and prosperity seemed assured, as the deprivations of wartime melted away in the expansive postwar climate. Plans big and small were made not only to recover ground lost during the war but to expand psychology's academic territory.

9

Remodeling the Academic Home

In May 1945, as the war was entering its final stages, Harvard president James Bryant Conant appointed a blue-ribbon panel to advise him on how to cope with the chaotic situation in psychology at the university. Like a set of competing fiefdoms, Harvard psychology was divided among several powerful figures and scattered across a number of administrative units. Christened the University Commission to Advise on the Future of Psychology at Harvard, the twelve-member group was headed by Rockefeller Foundation executive Alan Gregg, a physician, and included a number of prominent educators and psychologists. Rather than focus exclusively on the Harvard case, however, the group was charged with the broader task of determining "the place of psychology in an ideal university." That Harvard might approximate the ideal and thereby provide a blueprint for the rest of the academic world was an unspoken assumption.[1]

The commission labored for two years before publishing its report, *The Place of Psychology in an Ideal University*. The report contained few surprises, as it endorsed most of the changes that were already occurring in the postwar expansion of psychology. The group affirmed the notion that psychology was valuable for a wide variety of other fields and should be incorporated into professional training for education, medicine, business, law, theology, and engineering. It recommended that working contact between all psychologists on a campus be maintained and that training programs for applied psychologists be developed further. The report expressed concern that nonexperimental areas be given proper support, arguing that "arrangements which encourage the exclusive domination of psychology by the laboratory would sacrifice the unity of the subject, belie its freedom, and limit

---

1 Conant had tried a similar approach two years earlier, when he appointed the Harvard Committee on the Objectives of a General Education in a Free Society to advise on possible undergraduate curricular reforms. See Report of the Harvard Committee, *General Education in a Free Society* (Cambridge, Mass.: Harvard University Press, 1945).

its opportunity." The commission also strongly favored the creation of a new graduate degree, the Doctor of Psychology, or Psy.D., to designate practitioners. But the recommendation was not unanimous, as two psychologists in the group, Ernest Hilgard and Walter Hunter, registered their objections.[2]

The "Gregg Report" provided a framework for ongoing discussions about academic organization.[3] Several commentators pointed out how psychology had flourished under different administrative arrangements around the country. Local preference and institutional history had shaped organizational structures more profoundly than abstract prescriptions or theoretical models. Madison Bentley, at Cornell, concluded his long and thoughtful review of the volume by expressing his desire to see how its "abstract recommendations" would be "translated into the various current dialects of clinic, counsellor's booth, testing office, laboratory, computing-room, prison-grill, health center, analyst's chamber, and other working compartments actually housing its all-inclusive unit-for-psychologists."[4]

These comments accurately captured the tension inherent in the Harvard report between the abstract vision of psychology as a unified field and the working reality that psychology was fragmented along almost any conceivable dimension – intellectual, organizational, or occupational. The fact that psychologists could be found in any number of different departments in a single university, at Harvard and elsewhere, was a reflection of the field's polymorphous nature. Such diversity, however deplorable in the abstract, was the secret strength of the field, as it could present itself as a jack-of-all-trades science more successfully than most disciplines.

The cosmic tone of the commission's mandate, however, could not disguise the fact that it represented an attempt by Conant to deal with some disturbing local developments. For years the psychology department had been a bastion of experimentalism under the leadership of Edwin Boring, and those interested in clinical, social, and related fields had had to cope with second-class status. Psychometrics and mental testing were likewise relegated to the margins in the education school. During the war, however, the contributions of such fields were highlighted, and the balance of power began to shift at Harvard as well as at other institutions. As the rift between the "biotropic" and "sociotropic" psychologists at Harvard widened, mem-

2 Alan Gregg, Chester I. Barnard, Leonard Carmichael, Thomas M. French, Walter S. Hunter, Louis L. Thurstone, Detlev W. Bronk, John Dollard, Ernest R. Hilgard, Edward L. Thorndike, John C. Whitehorn, and Robert M. Yerkes, *The Place of Psychology in an Ideal University: The Report of the University Commission to Advise on the Future of Psychology at Harvard* (Cambridge, Mass.: Harvard University, 1947).

3 See Dael Wolfle, "The Place of Psychology in an Ideal University," *American Psychologist,* 1948, *3:*61–64; Walter V. Bingham, "Special Review: Psychology in an Ideal University," *Journal of Applied Psychology,* 1948, *32:*321–324.

4 Madison Bentley, "The Harvard Case for Psychology," *American Journal of Psychology,* 1948, *61:*275–282, quoted on p. 282.

bers of the latter group were threatening to join forces with sympathetic colleagues in sociology and anthropology as early as 1943.[5]

The disaffection of the sociotropes in psychology meshed with the ambitions of sociologist Talcott Parsons, who spearheaded the drive for a new interdisciplinary social science organization at Harvard. In 1946 a new Department of Social Relations was formed, headed by Parsons. It incorporated the fields of sociology, cultural anthropology, social psychology, and clinical psychology, and included a Laboratory of Social Relations directed by Samuel Stouffer. The fission between the sociotropic and biotropic psychologists was nearly complete, as Allport, Murray, and their students joined the new Department of Social Relations, while Boring, Stevens, Newman, and their experimentalist colleagues and students remained in the Department of Psychology.[6]

Therefore, when the *The Place of Psychology in an Ideal University* appeared in 1947, it was a dead letter, at least in regard to local developments. Ironically, it called for the integration of all psychologists into a single comprehensive department. That point was moot, however, at least at Harvard, as psychologists were now divided among two major departments and several other academic units.[7]

The Massachusetts Institute of Technology

On the other side of Cambridge, Massachusetts, less grandiose plans for refurbishing psychology's academic home were being made. During the latter part of the war, MIT officials negotiated with émigré German psychologist Kurt Lewin to create a new psychological research institute. The Research Center for Group Dynamics was established in 1945, bringing Lewin along with several coworkers from the University of Iowa. The center was dedicated to the exploration of the field of group dynamics, Lewin's term for the study of personality in relation to social psychology. Its members also shared the conviction that theory and practice were not only inseparable but essentially reinforcing to each other; this integrated approach to psychology they called "action research." Only two years after it opened, the Research Center for Group Dynamics lost its leader when Lewin died

5 G. W. Allport, C. K. M. Kluckhohn, O. H. Mowrer, H. A. Murray, and T. Parsons to P. H. Buck, 1 September 1943; Harvard University Archives/Harvard Psychology Department Papers/Box 1/Folder "Buck 1943–44/2."

6 Gordon W. Allport and Edwin G. Boring, "Psychology and Social Relations at Harvard University," *American Psychologist,* 1946, *1:*119–122; S. S. Stevens and E. G. Boring, "The New Harvard Psychological Laboratories," *American Psychologist,* 1947, *2:*239–243. See also Talcott Parsons, *Department and Laboratory of Social Relations, Harvard University: The First Decade, 1946–1956* (Cambridge, Mass.: Harvard University, 1957).

7 In the foreword, President Conant remarked: "Those familiar with Harvard and its tradition of dissent will not be altogether surprised and may even be amused that affairs should have taken this somewhat unusual turn."

at the age of fifty-seven. Its programs continued, however, even after it moved from MIT to the University of Michigan in 1947, where it became part of a major academic complex.[8]

During the war Lewin published extensively, on both programmatic issues in social psychology and empirical wartime research on American food habits. He was frequently a consultant to various government agencies and was involved in a number of interdisciplinary applied research projects, many with anthropologist Margaret Mead. As his thoughts turned increasingly toward the problems of making psychology a practical science without sacrificing analytical rigor, he reflected on the evolution of the field and in 1944 wrote:

> The relation between scientific psychology and life shows a peculiar ambivalence. In its first steps as an experimental science, psychology was dominated by the desire of exactness and a feeling of insecurity. Experimentation was devoted mainly to problems of sensory perception and memory, partly because they could be investigated through setups where the experimental control and precision could be secured with the accepted tools of the physical laboratory. As the experimental procedure expanded to other sections of psychology and as psychological problems were accepted by the fellow scientist as proper objects for experimentation, the period of "brass instrument psychology" slowly faded. Gradually experimental psychology became more psychological and came closer to life problems, particularly in the field of motivation and child psychology.
>
> At the same time a countercurrent was observable. The term "applied psychology" became – correctly or incorrectly – identified with a procedure that was scientifically blind even if it happened to be of practical value. As the result, "scientific" psychology that was interested in theory tried increasingly to stay away from a too close relation to life.[9]

Having to deal with "natural" (i.e., groups that they did not define) and practical problems in the context of war forced scientific psychologists to confront real-life situations, which Lewin viewed as a great opportunity. It was in this wartime context that he argued for an even closer and more symbiotic relationship between theory and practice in psychology:

> The greatest handicap of applied psychology has been the fact that, without proper theoretical help, it had to follow the costly, inefficient, and limited

8 For biographical detail, see Alfred J. Marrow, *The Practical Theorist: The Life and Work of Kurt Lewin* (New York: Basic Books, 1969).

9 Kurt Lewin, "Constructs in Psychology and Psychological Ecology," *University of Iowa Studies in Child Welfare,* 1944, *20:*23–27; reprinted as "Problems of Research in Social Psychology," in Dorwin Cartwright, ed., *Field Theory in Social Science: Selected Theoretical Papers* (New York: Harper & Row, 1951), pp. 155–169, quote on pp. 168–169.

method of trial and error. Many psychologists working today in an applied field are keenly aware of the need for close cooperation between theoretical and applied psychology. This can be accomplished in psychology, as it has been accomplished in physics, if the theorist does not look toward applied problems with highbrow aversion or with a fear of social problems, and if the applied psychologist realizes that there is nothing so practical as a good theory.[10]

Interestingly, after the war only the latter part of Lewin's formula – "there is nothing so practical as a good theory" – was picked up and its reciprocal antecedent dropped, suggesting that, despite the increased status of applied work, theory retained its primacy.

As Lewin expanded his professional networks during the war, he explored the possibility of moving from Iowa. Overtures from Harvard, California, and MIT were considered, but in the end only the MIT negotiations bore fruit. One of Lewin's key allies at MIT was Douglas McGregor, a professor in the Industrial Relations Section. Trained as a psychologist, he was alert to the possibilities of applying social science to industrial management and was particularly intrigued by some of the experiments Lewin and his group had conducted in industry. McGregor played a key role in 1944 in convincing MIT officials that the proposed center would fit the engineering mission of the school. He argued that "there has been an increasing realization in recent years that the *social* problems of industry and of our whole society are closely linked to technological progress." In his view, the work in "social engineering" that Lewin and others were pursuing could prove as important as developments in aerodynamics or radar.[11] Working behind the scenes, Gordon Allport at Harvard added his strong endorsement to hiring Lewin at MIT, calling him "the most original figure in psychology next to Freud during the past fifty years."[12]

The fact that the research center would bring with it substantial extramural funding no doubt made the prospect more enticing to the MIT administration. Lewin had obtained a total of $280,000 in research grants: $25,000 a year for ten years from the American Jewish Congress and $15,000 a year for two years from the Field Foundation. Only Lewin's salary would be paid by MIT; the rest of the staff payroll would be charged to the grants. Thus the research center represented

10 Cartwright, *Field Theory in Social Science,* p. 169.
11 Douglas R. McGregor (memo re proposed Research Center on Group Dynamics), September 1944; MIT Archives/AC4/32/9.
12 Allport had been part of a failed effort to create a position for Lewin at Harvard. R. G. C[aldwell], Dean of Humanities, "Memorandum Re: Research Center on Group Dynamics," 8 September 1944; MIT Archives/AC4/32/9.

a relatively low-cost investment that might reap large dividends. Lewin accepted the position in November 1944 and began working in March 1945.[13]

Initial steps to organize the center were undertaken over the next several months. With the war still on, Lewin could not immediately fill all of the staff positions he felt necessary. The research program revolved around the work of Marian Radke, whom Lewin brought with him from Iowa, and the field of minority problems. The Research Center for Group Dynamics made contacts with a variety of organizations to serve as "field cooperators" where observations and experiments could be carried out. Some courses were given, and a Ph.D. program in group psychology was approved. Lewin also spent time publicizing the new center. In the May 1945 issue of *Sociometry,* he described the new enterprise and justified its existence at MIT:

> The link between engineering and the total culture of a people has become more obvious as engineering undertakings have grown to gigantic dimensions. . . . The TVA or any one of the large river projects makes it apparent that modern culture has become so much saturated with engineering that the engineer can not help influencing deeply every aspect of group life by his action and by his omissions. It seems, therefore, entirely appropriate for an engineering school which perceives the main task of engineering in a progressive spirit, to make the scientific study of group life a part of its undertaking.[14]

By December 1945, with the war over, Lewin had assembled the core of his staff, including Marian Radke, Leon Festinger, Dorwin Cartwright, and others, as well as a handful of graduate students.[15]

One of Lewin's last articles was "Frontiers in Group Dynamics," published in 1947. At the beginning he argued that the war had accelerated the development of the social sciences, bringing changes potentially "as revolutionary as the atom bomb." The great advances in applied techniques, however, needed to be matched by equal progress in theory so that society could win "the race against the destructive capacities set free by man's use of the natural sciences." Lewin saw three main objectives: to integrate the social sciences; to move "from the description of social bodies to dynamic problems of changing group life"; and to develop new research instruments and techniques.[16]

---

13 Caldwell to Lewin, 8 September 1944; Lewin to Killian, 9 November 1944; MIT Archives/ AC4/32/9.

14 Kurt Lewin, "The Research Center for Group Dynamics at the Massachusetts Institute of Technology," *Sociometry,* 1945, *8:*126–136, on p. 134.

15 Kurt Lewin, "Research Center for Group Dynamics, The First Two Terms," 5 December 1945; MIT Archives/AC4/33/14.

16 Kurt Lewin, "Frontiers in Group Dynamics," *Human Relations,* 1947, *1:*143–153; reprinted in *Field Theory in Social Science,* pp. 188–237, quote on pp. 188, 189.

Lewin invoked the familiar arguments of cultural lag in order to support his attempts to quantify and mathematize the social constraints on individual behavior for utilitarian purposes. In the end, the intellectual constructs of action research and group dynamics and the institutional vehicles devoted to their elaboration and propagation displayed Lewin's intuitive grasp of the research laboratory as a source of professional power. His own career testified to the truth of his trademark slogan that "there is nothing so practical as a good theory."

Lewin's untimely death in early February 1947 left the fate of the Research Center for Group Dynamics in question. Without Lewin, MIT administrators were skeptical about the center's future, and by early March had decided to terminate it by the end of the following academic year.[17] The center and the bulk of its staff eventually moved to the University of Michigan, where Lewin's student Dorwin Cartwright took over the reins. At MIT, psychology reverted to its former status as an auxiliary discipline that provided service courses for science and engineering majors. Its importance was noted in various plans to reorganize the social sciences and humanities programs, and several psychologists were hired by various units, but it retained its service role into the early 1950s.

By 1953 there were thirteen psychologists in the MIT faculty ranks, including one with permanent tenure. Among them were associate professor George A. Miller, a veteran of the Harvard Psycho-Acoustic Laboratory and rising star in the psychology of language and communication. The presence of Miller and others with strong reputations in research led to efforts to create an independent psychology department under the School of Science. In 1953 an outside review committee composed of Donald Marquis of Michigan, Lyle Lanier of Illinois, and Smith Stevens of Harvard reported on the status and prospects of psychology at the institute. They suggested that MIT was poised to develop an outstanding program in engineering psychology, the kind of experimental psychology that "descended from the 'brass instrument' psychology of the older days and has become the 'black box' psychology of today." While accepting the traditional role granted to psychology in humanizing and socializing young engineers, this group of ambitious and research-minded scientific entrepreneurs wanted to integrate psychology with engineering through a focus on common technical goals. As Miller put it, such an emphasis meant the "generalization of the scientific method to include people as well as machines."[18]

17 J. R. Killian, Jr., "Memorandum of conversation with Prof. Douglas R. McGregor," 8 March 1947; MIT Archives/AC4/55/2.

18 D. G. Marquis, L. H. Lanier, and S. S. Stevens, "Psychology at MIT," 3 October 1953; Donald G. Marquis Papers/MIT Archives/AC20/Folder 120; George A. Miller, "A Survey of Psychology in Schools of Engineering," 1 July 1952; MIT Archives/AC20/3/119.

The University of Chicago

One of the most dramatic revivals of academic psychology after the war occurred at the University of Chicago. In 1947 James Grier Miller, head of the Veterans Administration clinical psychology program, accepted the chairmanship of the practically moribund Department of Psychology at Chicago. Under the administration of university president Robert Hutchins since 1929, the number of faculty members in psychology had dwindled, and Chicago fell from its position as a leading producer of Ph.D.s. Miller, with his cosmopolitan Harvard background, was apparently able to convince Hutchins that his humanistic university might profit from a fresh transfusion of scientific psychologists.[19]

Miller, who counted both Edwin Boring and Henry Murray among his mentors in psychology, was interested in implementing the Harvard Commission's recommendations in the *Place of Psychology in an Ideal University* when he went to Chicago. A friend of Alan Gregg, who had chaired the group, Miller had some affinity for the medical model of professionalization, with its bifurcation of clinical practitioners and clinical researchers. (He himself held an M.D. as well as a Ph.D. in psychology.) For psychology, however, he clearly favored an emphasis on the development of clinical research as a foundation for practice and apparently never seriously pursued the idea of the Psy.D. as a separate professional degree at Chicago.

Miller assumed his duties in 1948 and quickly expanded the faculty ranks, increasing the staff from seven members to more than thirty by 1954. A rapid rise in doctorate production ensued. Between 1940 and 1948 only ten Ph.D.s had been awarded; in 1949 alone there were eleven, and the numbers continued to mount.[20]

Before long, Miller's initial ambitions for psychology at Chicago became subsumed within an even larger project, the development of the "behavioral sciences." In 1949 a group of senior faculty from the Social Sciences Division (under Dean Ralph Tyler) and the Biological Sciences Division (under Dean Lowell Coggeshall) were brought together to form the Committee on the Behavioral Sciences. It was chaired by Miller, who credited his physicist colleague Enrico Fermi with advancing the idea of an interdisciplinary effort to pursue a general theory of human behavior.[21]

Miller had a deep-rooted interest in the philosophy of the life sciences that was

19 Personal interview with James Grier Miller, 17 December 1994.
20 Totals for following years: 1950 (16); 1951 (28); 1952 (18); 1953 (13). "Ph.D.s in Psychology, 1903–48," JGM/1947–49/1949 Correspondence; "Doctoral Degrees Granted by the Psychology Department of the University of Chicago," JGM/1954/Dept. Psychology – Univ. Chicago; James G. Miller, "The Department of Psychology and the Behavioral Sciences," July 1954; JGM/1950–51/Misc.
21 "The Scientific and Professional Activities of James Grier Miller" [1993]; JGM.

first nurtured by contact with Alfred North Whitehead as a Harvard student in the 1930s, when the senior philosopher's architectonic vision inspired the budding psychologist. In the expansive postwar context, with psychologists embracing interdisciplinary and cooperative research, the time seemed ripe for such grand theorizing. After a couple of years of planning and preliminary meetings, a theory group began to meet on a weekly basis to explore the application of concepts from systems theory, cybernetics, and similar areas to the formation of an integrated theory. As Miller later described it, the "group had analyzed such issues as the unity of science, the various jargons of different disciplines and academic schools, reductionism, and formal identities and analogies among various types and levels of living entities."[22]

In 1953 Miller and the Committee on the Behavioral Sciences began to consider a permanent establishment. The university made it clear that the venture would have to depend on outside funding. Miller incorporated the Fund for the Behavioral Sciences as a nonprofit corporation in Illinois and enlisted his friend, Donald Marquis, chair of psychology at the University of Michigan, and David Shakow, his Chicago colleague, to serve as directors along with himself. To establish such an interdisciplinary institute would take an estimated $400,000 a year for the first decade.[23]

The Ford Foundation was the obvious target as a potential patron. With the help of Marquis as a consultant, the foundation had identified the behavioral sciences as an area for strategic investment and was beginning to make substantial grants in the area. Several universities, including Chicago, were given seed money to develop their programs and plans. After the creation of the Center for Advanced Study in the Behavioral Sciences was announced in 1952, Ralph Tyler, Dean of Social Sciences at Chicago, was chosen as its first director. Eventually stymied in his efforts to interest the Ford philanthropy in his plans, Miller turned to his old friend and colleague, Marquis at Michigan.

The University of Michigan

Perhaps the most successful attempt to capitalize on psychology's protean identity was made at the University of Michigan, where the field's multiple connections to biological, medical, and social science were exploited after the war. Under the leadership of Donald Marquis, the university became a major center for research and graduate education in psychology and exemplified the kinds of opportunities that enabled psychologists to expand their academic domain.

In 1942 Walter Pillsbury retired as head of the Department of Psychology at the

---

22 Ibid., p. 8. See also Steve J. Heims, *The Cybernetics Group* (Cambridge, Mass.: MIT Press, 1991).

23 The Committee on the Behavioral Sciences, University of Chicago, "Program of Research in the Behavioral Sciences" (c. 1953); JGM/1950–51.

University of Michigan after forty-five years of service. The dean, concerned about the effects of the department's long tradition of hiring its own doctorates to the faculty, decided to look outside for new leadership.[24] Conducted in the midst of the war, the search narrowed to several younger psychologists who were active in the war effort. Among the candidates considered were Ernest Hilgard, Clarence Graham, Norman Cameron, B. F. Skinner, E. G. Wever, and Donald Marquis.[25]

Marquis, whose name had been suggested by Leonard Carmichael, was chairman of the Yale psychology department and director of the wartime Office of Psychological Personnel. He visited the Michigan campus in the summer of 1944 and was subsequently hired to begin his duties in the fall of 1945. By virtue of his temperament and his training Marquis was uniquely suited to the task.

Although he had done significant experimental research early in his career, Marquis did not remain at the laboratory bench for long after completing his Ph.D. at Yale in 1932. He found writing agonizingly difficult and was much more comfortable in the classroom, at the conference table, and on the telephone. A leader among the junior faculty, he rose quickly through the ranks at Yale to become chairman of the psychology department in 1942. In addition to keeping the Yale program running under wartime conditions, Marquis accepted an assignment in 1943 in Washington to oversee the Office of Psychological Personnel. This small but vital operation served as an employment office and information center for the profession, and Marquis was able to extend his already wide contacts among his colleagues. He could also observe at first hand the development of new working relationships between scientific groups and the federal government.

At Michigan, Marquis was in a position to exercise his talents as an academic entrepreneur. Over a period of several years he was able to build a research empire, with the psychology department at the center of a network of programs extending into mental health, education, social research, and other areas. A master at tapping into extramural funding sources, he "put together half-salaries, research grants, VA stipends, training grants, industrial contracts" in order to stretch university funds.[26] Joint appointments between psychology and other university units helped to expand the staff and extend the reach of psychologists throughout the

24 Although the practice of hiring one's own graduates was common among psychology departments before World War II, "inbreeding" was especially pronounced at Michigan, where all but six of the twenty-six faculty appointed to full-time positions since the 1890s had been Michigan-trained. Alfred C. Raphelson, *Psychology at the University of Michigan: 1852–1950: Volume I, The History of the Department of Psychology* (Flint: University of Michigan Flint College, 1968), p. 67. See also Alfred C. Raphelson, "Psychology at Michigan: The Pillsbury Years, 1897–1947," *Journal of the History of the Behavioral Sciences,* 1980, *16:*301–312.

25 Raphelson, *Psychology at the University of Michigan,* pp. 69–75.

26 Robert R. Sears, "Donald George Marquis: 1908–1973," *American Journal of Psychology,* 1973, *86:*661–663, quote on p. 662.

university. Initially, arrangements were made with the School of Education, the Department of Sociology, the Psychological Clinic, the Bureau of Psychological Services, and the Counseling Bureau. By 1965 members of the psychology department held joint appointments with at least twenty-five other campus units.[27]

Michigan also welcomed self-financing institutes that associated with the department. In 1948 the Research Center for Group Dynamics relocated from MIT, and the Survey Research Center was established. The two units were joined administratively in 1949 under the umbrella of the Institute of Social Research. Again, many of the staff members received joint appointments in the Department of Psychology.[28]

Between 1945 and 1950 the department's faculty grew fivefold, from eight to forty members. Within a lag of a few years, Ph.D. production exhibited a correspondingly dramatic leap, increasing by an order of magnitude from the 1940s to the 1950s.[29]

Period	Number of Ph.D.s
1930–34	11
1935–39	15
1940–44	17
1945–49	10
1950–54	127
1955–59	126
1960–64	165
1965–69	146

Social psychology was perhaps the most impressive area of development. Under the direction of Theodore Newcomb, a doctoral program in social psychology was set up in 1947; over the next twenty years it produced a total of 226 Ph.D.s, helping to populate a growing field.[30]

In 1955 yet another major component was added to the growing Michigan empire. The Mental Health Research Institute was established with over $2.5 million in state and federal appropriations. It was headed by James Grier Miller, who

27 Raphelson, *Psychology at the University of Michigan,* pp. 78–79.
28 Dorwin Cartwright, "Some Things Learned: An Evaluative History of the Research Center for Group Dynamics," *Journal of Social Issues,* 1958, *Supplement Series,* no. 12: 1–19; Rensis Likert and Stanley E. Seashore, *Two Papers Describing the Origins, History and Present Organizational Life of the Institute* (Ann Arbor: University of Michigan Institute for Social Research, 1965); Charles F. Cannell and Robert L. Kahn, "Some Factors in the Origins and Development of the Institute for Social Research, University of Michigan," *American Psychologist,* 1984, *39:*1256–1266.
29 K. F. Riegel, "A Structural, Developmental Analysis of the Department of Psychology at the University of Michigan," *Human Development,* 1970, *13:*269–279, on p. 271.
30 Daniel Katz, "Theodore M. Newcomb: 1903–1984," *American Journal of Psychology,* 1986, *99:*293–298.

moved from the University of Chicago along with a core group of colleagues interested in behavioral science theory and research.[31]

Although he was much more active as an administrator than as a researcher, Marquis served as president of the American Psychological Association in 1948. He was the first of a half-dozen veterans of the Yale Institute of Human Relations elected to that office during the next several years. In his presidential address, "Research Planning at the Frontiers of Science," he generalized on the basis of his personal knowledge some principles involved in the management of large-scale cooperative research programs.[32] Recognized as a master at institution building, Marquis epitomized the emergence of the scientist-entrepreneur in postwar psychology.

The Johns Hopkins University

Different opportunities were available at The Johns Hopkins University, where the psychology department had been abolished in 1941. It was reestablished in 1946 with Clifford T. Morgan as chairman. Only thirty-one years old, Morgan had been tapped to build a new program in experimental psychology, both pure and applied, from the ground up.[33] Since his days as a graduate student under Leonard Carmichael at the University of Rochester, where he obtained his Ph.D. in 1939, Morgan was a "crusader" for physiological psychology. In 1939 he moved to Harvard on a temporary instructorship and became associated with Karl Lashley's laboratory, where he continued his research on the effects of insulin on food intake in white rats. A brilliant and magnetic teacher, he produced the classic textbook *Physiological Psychology* (1943) during his Harvard stay. His professional visibility increased even further as a result of his prominent role in a heated controversy over the cause of audiogenic seizures in the rat, which he argued were due to simple sensory stimulation rather than complex choice discriminations.[34]

In 1943 Morgan was hired as an assistant professor of biology at Hopkins. At the same time he became a technical consultant to Division 17 (Physics) of the Office of Scientific Research and Development. Among the projects he dealt with

31 See annual reports, Mental Health Research Institute, University of Michigan, 1957–1966.

32 Donald G. Marquis, "Research Planning at the Frontiers of Science," *American Psychologist,* 1948, *3:*430–438.

33 On the history of the Hopkins department and the events leading up to its demise, see Philip J. Pauly, "G. Stanley Hall and His Successors: A History of the First Half-Century of Psychology at Johns Hopkins," in Stuart H. Hulse and Bert F. Green, eds., *One Hundred Years of Psychological Research in America: G. Stanley Hall and the Johns Hopkins Tradition* (Baltimore: Johns Hopkins University Press, 1986), pp. 21–51.

34 Eliot Stellar and Gardner Lindzey, "Clifford T. Morgan: 1915–1976," *American Journal of Psychology,* 1978, *91:*343–348.

was the Psycho-Acoustic Laboratory at Harvard, headed by his close professional associate Smith Stevens. After the war, Morgan recruited a large group of wartime psychology veterans to continue their studies of man–machine systems in the Institute of Cooperative Research, located off-campus. This group formed the core of a new department of psychology. Composed almost entirely of junior professors, one faculty member remembered it as "a brash and even cocky department, intensely competitive and ambitious in Cliff's image."[35] The faculty was an all-male group, and women were not encouraged to apply to the graduate program.[36]

In 1949 three members of the faculty published a textbook entitled *Applied Experimental Psychology: Human Factors in Engineering Design*. It represented an effort to show the military usefulness of certain kinds of precise laboratory research and demonstrated how rigor and relevance in psychology were welded together by the war.[37] Morgan went on to head the Systems Development Laboratory at Hopkins, which was dedicated to the design of sophisticated control systems for air traffic and to research in other areas that had overlapping civilian and military applications.[38]

The Expansion of Undergraduate Education

Psychology departments, like colleges and universities more generally, benefited enormously from the postwar boom in American higher education. The G.I. Bill, a robust economy, and a widespread consensus on the positive value of a college degree helped usher in a "golden age" of expansion. The growth of academic psychology was built on a broadening foundation of undergraduate education. If graduate training was the place where psychologists came to grips with the pressures of intellectual fragmentation and occupational specialization, then undergraduate education was where a more unitary vision of psychology held sway. For aspiring young psychologists the problems of professionalization could be deferred until they entered graduate training. The introductory course thus acted as a centripetal force for the indoctrination of students into the prevailing orthodoxy of scientific professionalism.

In the postwar decades psychology proved to be an enormously popular undergraduate subject. As university enrollments burgeoned, a robust market developed

---

35 Stellar, in ibid., p. 345.
36 Pauly, "G. Stanley Hall and His Successors," pp. 41–42. Among the assistant professors were Neil Bartlett, Alphonse Chapanis, Wendell Garner, Eliot Stellar, and Stanley Williams.
37 Alphonse Chapanis, Wendell R. Garner, and Clifford T. Morgan, *Applied Experimental Psychology: Human Factors in Engineering Design* (New York: Wiley, 1949).
38 See W. R. Garner, "Clifford Thomas Morgan: Psychonomic Society's First Chairman," *Bulletin of the Psychonomic Society,* 1976, 8:409–415.

for introductory textbooks. The genre took a new turn as some postwar textbooks took a fresh approach to the diversity of psychological science.[39]

The tension between presenting psychology as a rigorous science and showing its relevance to the concerns of the average student had been a long-standing problem in the writing of undergraduate textbooks. In the 1930s there were several competing books that took different approaches. Among them were: Robert S. Woodworth's *Experimental Psychology* (1938), which provided an acknowledged vade mecum that was probably more suited to the needs of graduate students; Boring, Langfeld, and Weld's *Psychology: A Factual Textbook* (1935), which emphasized traditional experimental topics organized under conventional rubrics such as sensation and perception; and Floyd Ruch's *Psychology and Life* (1937), which attempted to convey psychological knowledge in relation to the issues and concerns of students. All of these approaches had their drawbacks. The tradeoff seemed to be between explaining psychology adequately from a scientific point of view and maintaining student interest in the subject.[40]

Among those who tried to attract students while retaining the respect of their scientific colleagues was Norman L. Munn, a professor at Vanderbilt University and veteran textbook author. His first book was a handbook on animal behavior that provided a comprehensive survey of research concerning the white rat. Published in 1933, *An Introduction to Animal Psychology: The Behavior of the Rat* was produced with the support of Walter Hunter, editor of *Psychological Abstracts*. A few years later Munn wrote an introductory textbook on genetic (i.e., developmental) psychology.[41] Having discovered his métier, in 1940 he began working on a general introduction to psychology, hoping to overcome the deficiencies of existing textbooks.

Munn wanted the book to "paint a picture of man as the psychologist views him and show the application of psychological methods, principles and facts to the student's understanding of his own behaviour and experience and to the solution of the everyday problems of the world we live in." Such an approach meant that the text would have to be organized eclectically instead of by the traditional "schools and systems" format.[42] Munn's progress on the book was halted by Pearl Harbor,

39 For a discussion of the functions of introductory textbooks with examples from the early twentieth century, see Jill G. Morawaki, "There Is More to Our History of Giving: The Place of Introductory Textbooks in American Psychology," *American Psychologist,* 1992, *47:*161–169.

40 Robert S. Woodworth, *Experimental Psychology* (New York: Holt, 1938); E. G. Boring, H. S. Langfeld, and H. P. Weld, eds., *Psychology: A Factual Textbook* (New York: Wiley, 1935); Floyd L. Ruch, *Psychology and Life* (Glenview, Ill.: Scott, Foresman, 1937).

41 Norman L. Munn, *An Introduction to Animal Psychology: The Behavior of the Rat* (Boston: Houghton Mifflin, 1933); idem, *Psychological Development: An Introduction to Genetic Psychology* (Boston: Houghton Mifflin, 1938).

42 Norman L. Munn, *Being and Becoming: An Autobiography* (Adelaide, South Australia: Adelaide University Union Press, 1980), p. 104.

and he put the manuscript away for nearly three years before returning to it in 1945.

Psychology: The Fundamentals of Human Adjustment came off the press in April 1946, one of the first new textbooks to hit the postwar market. Although the title stressed human psychology, the author's acquaintanceship with the findings of animal psychology was clear. In adhering to his aims, Munn emphasized established facts and downplayed theoretical controversy. In the preface, he urged students to "look upon this as a book about yourself."[43] The volume was heavily illustrated, and the text was laid out in double columns to conserve paper in the face of continuing shortages. The cover featured a circular maze pattern, reminiscent of the ubiquitous rat research tool as well as traditional garden-court mazes. The book was highly successful from the start. Adopted for course use by some five hundred colleges and universities, it sold more than 150,000 copies in its first year.[44]

In 1948 the venerable trio of Edwin Boring, Herbert Langfeld, and Harry Weld produced a third revision of their introductory textbook, first published in 1935. All three had been stalwart members of the prewar scientific establishment associated with traditional centers of experimentalism at Harvard, Princeton, and Cornell, respectively. Their text, entitled *Foundations of Psychology,* represented an attempt to capture part of the postwar market with an updated version of experimentalism.[45]

As before, Boring dominated the collaboration. The book's tone and coverage reflected his own evolving sense of the proper definition of psychological science. Unlike the two previous editions, *Foundations of Psychology* incorporated significant amounts of material on what Boring termed "sociotropic" psychology (i.e., social, clinical, personality, and applied psychology more generally). The balance between the coverage of "biotropic" psychology (i.e., sensation, perception, learning, and experimental psychology more generally) and sociotropic psychology put the book more in line with the mainstream.

In fact, nearly half of the 1948 text dealt with sociotropic topics, compared to less than 20% in the 1935 text. This striking change can be attributed to Boring's conversion to the cause of applied psychology during the war, which was exhibited in his authorship of *Psychology for the Fighting Man* (1943). That book, along with a West Point textbook, *Psychology for the Armed Services* (1945), contains a preponderant emphasis on nonexperimental subjects, which reflects Boring's response to wartime demands for practical information. When all of Boring's textbooks are

---

43 Norman L. Munn, *Psychology: The Fundamentals of Human Adjustment* (Boston: Houghton Mifflin, 1946; 2d ed. 1951); quote from p. xi.

44 Munn, *Being and Becoming,* pp. 112–127.

45 Edwin Garrigues Boring, Herbert Sidney Langfeld, and Harry Porter Weld, *Psychology: A Factual Textbook* (New York: Wiley, 1935); idem, *Introduction to Psychology* (New York: Wiley, 1939); idem, *Foundations of Psychology* (New York: Wiley, 1948).

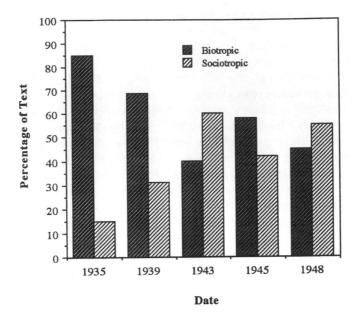

Figure 9.1. Biotropic vs. sociotropic emphasis in texts by E. G. Boring. *Sources:* Boring, Langfeld, and Weld, *Psychology: A Factual Textbook,* 1935; Boring, Langfeld, and Weld, *Introduction to Psychology,* 1939; Boring and Van de Water, *Psychology for the Fighting Man,* 1943; Boring, *Psychology for the Armed Services,* 1945; Boring, Langfeld, and Weld, *Foundations of Psychology,* 1948.

analyzed according to his own categories of biotropic and sociotropic emphasis, the trend toward the sociotropic is dramatic and unmistakable. As the lessons of war were assimilated into professional consciousness, the trend exemplified in the writing of Boring was more generally reflected in the textbook literature (Figure 9.1).[46]

Reacting to the robust demand, Munn produced a second edition of his textbook in 1951. Laying even greater stress on human psychology, this edition included results from research conducted during World War II. The contents were divided among the following section headings: the scope and method of psychology; the organism; learning; memory and thinking; motivation; emotion; "knowing our world" (sensation and perception); and individual differences. In its general form, this pattern of coverage became widely used in introductory textbooks. The second edition of *Psychology: Fundamentals of Human Adjustment* was even more

46 See Dael Wolfle, "The First Course in Psychology," *Psychological Bulletin,* 1942, *39:* 685–712; Ned Levine, Colin Worboys, and Martin Taylor, "Psychology and the 'Psychology' Textbook: A Social Demographic Study," *Human Relations,* 1973, *26:*467–478.

successful than the first. The first two editions sold a combined total of more than a million copies before they were superseded by the third edition in 1956.[47]

Paced by Munn's book, a new standard in the tone and coverage of undergraduate textbooks emerged rather quickly after the war as the move toward student-oriented texts accelerated.[48] In 1953 Stanford psychologist Ernest Hilgard published his *Introduction to Psychology,* which contained many of the same features as Munn's textbook. Hilgard's approach was more encyclopedic and comprehensive, and he provided extensive references to the research literature.[49] One innovation in format that was soon widely copied was the use of sidebars to break up the main mass of text and provide additional details on selected topics.

Psychologists also attempted to foster a market for high school level textbooks. For instance, in 1945 Thelburn L. Engle, a psychology professor at Indiana University's branch campus in Fort Wayne, published *Psychology: Its Principles and Applications* as an elementary textbook that could be used for either upper-level high school courses or lower-level college courses. Engle, a former high school teacher of mathematics and psychology, presented psychology as a scientific subject with important practical consequences. In addition to covering the scientific methods and findings of psychology, the text was designed to help students "develop what is best in their personalities" and to become responsible family and community members. Aside from a simpler writing style and lists of discussion questions, the volume was similar to contemporary college textbooks.[50]

The undergraduate textbook market remained large and relatively stable for decades. Lineages of many of the major textbooks of today go back thirty, forty, or even fifty years, suggesting how interest in a standardized undergraduate

---

47 Munn, *Being and Becoming,* p. 132. By 1985 Munn's textbook, with new authors, had reached its ninth edition; Wayne Weiten and Randall D. Wight, "Portraits of a Discipline: An Examination of Introductory Psychology Textbooks in America," in Antonio E. Puente, Janet R. Matthews, and Charles L. Brewer, eds., *Teaching Psychology in America: A History* (Washington, D.C.: American Psychological Association, 1992), pp. 453–504, on p. 467.

48 There was already some movement toward student-oriented texts, exemplified by Ruch's *Psychology and Life* (1937), which used the notion of personal adjustment as a motif. In the second edition (1941) Ruch responded to criticisms about the book's scientific content and included some traditional topics, such as brain and nervous system, that had not been included in the previous edition. By 1992 the textbook had reached its thirteenth edition; Weiten and Wight, "Portraits of a Discipline," pp. 465–466.

49 Ernest R. Hilgard, *Introduction to Psychology* (New York: Harcourt, Brace, 1953). Hilgard remained a coauthor until this spectacularly successful book reached its ninth edition in 1987; Weiten and Wight, pp. 467–468.

50 T. L. Engle, *Psychology: Its Principles and Applications* (Yonkers-on-Hudson, N.Y.: World Book, 1945; rev. ed. 1950); quote from rev. ed., p. iii.

curriculum was powerfully reinforced by the economics of publishing. The result has been a remarkably homogeneous and stable genre of introductory text-books.[51]

In keeping with its activist postwar thrust, the American Psychological Association became involved in undergraduate curriculum review and reform. APA Executive Secretary Dael Wolfle convened a small group of academic psychologists in 1951 to conduct an "audit" of undergraduate instruction. The group cited four possible educational objectives:

> (1) Intellectual development and a liberal education; (2) a knowledge of psychology, its research findings, its major problems, its theoretical integrations, and its contributions; (3) personal growth and an increased ability to meet personal and social adjustment problems adequately; (4) desirable attitudes and habits of thought, such as the stimulation of intellectual curiosity, respect for others, and a feeling of social responsibility.[52]

Most concerned about increasing scientific literacy, the group declined to endorse the goal of personal growth and self-knowledge in its recommendations, preferring instead to emphasize the mastery of psychological knowledge and scientific reasoning.

A decade later another group met to revisit the issues, with the financial support of the National Science Foundation. By that time increasing enrollments had put additional strains on the teaching mission of psychology departments, and there was concern in some quarters that undergraduate instruction was not receiving the attention it deserved in a research-oriented reward system. In contrast to the earlier group's recommended ideal curriculum, this group proposed a set of model curricula matched to the needs of different kinds of institutions and underscored the traditional commitment to psychology in the context of the liberal arts. Although the report tried to avoid excessive vocationalism, it was clear that the era of the "academic shopping center" in undergraduate psychology had arrived.[53]

51 See Weitan and Wight, "Portraits of a Discipline."
52 C. E. Buxton, C. N. Cofer, J. W. Gustad, R. B. MacLeod, W. J. McKeachie, and D. Wolfle, *Improving Undergraduate Instruction in Psychology* (New York: Macmillan, 1952), pp. 2–3, quoted in Thomas V. McGovern, "Evolution of Undergraduate Curricula in Psychology, 1892–1992," in Puente et al., *Teaching Psychology in America*, p. 28. See also Margaret A. Lloyd and Charles L. Brewer, "National Conferences on Undergraduate Psychology," in ibid., pp. 263–284.
53 W. J. McKeachie and J. E. Milholland, *Undergraduate Curricula in Psychology* (Glenview, Ill.: Scott, Foresman, 1961); McGovern, "Evolution of Undergraduate Curricula," pp. 28–29.

Reconsidering Graduate Education

The challenges posed by continued professional growth and accelerating scientific change in psychology converged in graduate education. Here the process of producing new knowledge was intimately joined to the process of producing new knowers. Specialization, occupational as well as intellectual, was an unavoidable feature of professional life, and various accommodating mechanisms had evolved in response to its manifold challenges. In the process of training new psychologists issues of scientific coherence and professional unity had to be faced squarely. What knowledge should all psychologists possess? What attitudes and values should they bring to their work? In practical terms, how should graduate education proceed?

These questions were perennial, but they achieved a new urgency by the middle of the 1950s as psychology rode a wave of growth that showed no sign of cresting. Following the landmark 1949 Boulder conference on training in clinical psychology were a series of conferences devoted to graduate education in other areas, including counseling psychology (1951), school psychology (1954), psychology and mental health (1955), and research training (1958).[54] In late 1958 the next conference was held at Miami, Florida, with 122 psychologists in attendance. This time the entire subject of graduate education was on the agenda.

By the time of the Miami conference the annual production of new doctorates in psychology was approaching eight hundred, nearly three times the number awarded in 1949 only a decade earlier.[55] There were more than two hundred American colleges and universities offering graduate work in psychology, nearly half of which were Ph.D.-granting institutions. Demand for professional psychologists was driven by a number of factors, including population growth, urbanization, the expansion of higher education, and shortages in other professions, especially psychiatry. The boom in nonacademic employment posed a particular challenge. As

---

54 For published reports on the respective conferences, see Victor C. Raimy, ed., *Training in Clinical Psychology* (New York: Prentice-Hall, 1950); Committee on Counselor Training, Division of Counseling and Guidance, American Psychological Association, "Recommended Standards for Training Counseling Psychologists at the Doctorate Level," *American Psychologist*, 1952, 7:175–181; Norma E. Cutts, ed., *School Psychologists at Mid-Century* (Washington, D.C.: American Psychological Association, 1955); C. R. Strother, ed., *Psychology and Mental Health* (Washington, D.C.: American Psychological Association, 1956); Donald W. Taylor, H. F. Hunt, and W. R. Garner, "Education for Research in Psychology: Report of a Seminar Sponsored by the E & T Board," *American Psychologist*, 1959, 14:167–179. For an overview, see Cynthia Belar, "Education and Training Conferences in Graduate Education," in Puente et al., *Teaching Psychology in America*, pp. 285–299.

55 Lindsay R. Harmon, *A Century of Doctorates: Data Analyses of Growth and Change* (Washington, D.C.: National Academy of Sciences, 1978), p. 13.

Joseph Bobbitt, Assistant Director of the National Institute of Mental Health, asserted in his opening remarks: "The educators of psychologists can no longer merely reproduce their own image or make separate arrangements for one difficult to understand sport in their ranks. Rather, the graduate department of today faces the fact that it is educating psychologists who will be doing many practical and remunerative things far from college halls."[56] Expanding employment opportunities in the mental health services were changing occupational realities.

At the conference it seemed a foregone conclusion that graduate programs would expand. The possibility of finding ways to curtail growth was apparently not even broached. Instead, participants sought ways to keep pace with emerging demands and to rationalize the expansion of graduate training. As they discussed the challenges of specialization, they acknowledged the need to harmonize the goals of professional service and the aims of scientific inquiry.

Much debate surrounded the issue of the existence of a common core of psychological knowledge and how it related to the graduate curriculum. Using the rhetoric of experimentation, some argued that the curriculum was an independent variable, and thus at least partly responsible for defining elements of a common core. Others held that the curriculum was a dependent variable that reflected some shared vision of basic psychological knowledge. All agreed that a common core existed, but the group declined to specify exactly what it entailed, ostensibly in order not to "discourage imaginative innovation in graduate training."[57] It also had the effect of avoiding a futile search for consensus.

To ensure some degree of conformity in the coverage of core topics the conference relied on disciplinary tradition as well as standardized assessments such as the Graduate Record Examination and the certification tests of American Board of Examiners in Professional Psychology. Training in research was held up as the other essential and distinctive feature of graduate education in psychology. Proposing no major departures, the Miami conference endorsed a "new gestalt" of the psychologist's role that embraced both the traditional "pure scientist" and the "scientist-practitioner" from the Boulder conference.[58]

This middle-of-the-road approach was reflected in the conclusion of the conference report, which ended on a strong note of affirmation: "a striking aspect of the Conference was the great cohesiveness demonstrated; the often expressed fears of imminent fractionation of psychology seem quite unfounded in December, 1958."[59]

Among the issues raised by the growing divergence between scientific and professional psychology was the advisability of separate graduate training tracks for

56 Joseph M. Bobbitt, "Opening Remarks," in Anne Roe, John W. Gustad, Bruce V. Moore, Sherman Ross, and Marie Skodak, eds., *Graduate Education in Psychology* (Washington, D.C.: American Psychological Association, 1959), pp. 19–23, quote on p. 21.
57 Ibid., p. 44.   58 Ibid., p. 42.   59 Ibid., p. 89.

research scientists and professional practitioners. Some questioned the dominant "scientist-practitioner" model of training endorsed by the Boulder conference in 1949. A new doctoral degree analogous to professional degrees in medicine, law, and other fields was proposed as a means of improving graduate training as well as a way to more clearly differentiate professional roles. In some quarters the idea of a Doctor of Psychology (Psy.D.) degree was raised again. (It had been suggested in the 1947 Harvard report on *The Place of Psychology in an Ideal University.*) But the Miami conference recorded its official "indifference" to the question of a separate professional doctoral degree.[60]

In discussing ways to improve the education of psychotherapists and clinicians, the participants in the APA conference on psychology and mental health held in 1955 emphasized the importance of practicum experiences that would expose students to real-life problems. This was in keeping with a general concern expressed by Carl Rogers and others that graduate training in clinical psychology should involve more than the formal transmission of knowledge. A few years later, at a parallel conference on education for research, participants came to a similar conclusion: "perhaps the essence of good research training is also experiential, rather than cognitive." This convergence of views was noted at the Miami conference in 1958 on graduate training in general.[61]

The Psychologist of the Future

Several months earlier, one of the participants in the Miami conference, Stuart W. Cook, chair of the New York University psychology department, expressed more pointed concern over the increasing divergence between the scientific and the professional roles of American psychologists. In his presidential address to the Eastern Psychological Association, Cook wondered what "the psychologist of the future" might look like, since "history has bequeathed us one type of psychology and society is insisting that we produce, in addition, a second type." He sketched three scenarios. In the first, the term "psychologist" would be reserved for those filling traditional academic positions as scholars and researchers, and a new label, such as "psychotechnologist," would be applied to those pursuing professional service roles. A second possibility was that "psychologist" would reflect common popular usage and refer to practitioners of all kinds, both academic and nonacademic. In this case, scientific investigators would be identified by their specialty area, such as animal behavior, sensory processes, or personality. The third scenario followed existing tradition and took "psychologist" as an inclusive term. In

60 Ibid., p. 63.
61 Ibid., p. 55. See also Strother, ed., *Psychology and Mental Health;* Taylor, Hunt, and Garner, "Education for Research in Psychology: Report of a Seminar Sponsored by the E & T Board."

this case, specialization would be accommodated in the later stages of graduate training, and no attempt would be made to make hard and fast distinctions among scientific and professional dimensions of the role. Like others, Cook saw the advantages of a productive symbiosis and the dangers of a split that might lead to "a commonplace profession and a disembodied science."[62]

62 Stuart W. Cook, "The Psychologist of the Future: Scientist, Professional, or Both?" *American Psychologist,* 1958, *13:*635–644, quote on pp. 639, 644.

Interlude IV

In April 1928, with the publication of *A History of Experimental Psychology* a year away, Edwin Boring wrote to Carl Murchison about his need for "facts concerning the scientific development of certain individuals" that could be obtained only from autobiography. Murchison, a psychologist at Clark University, was an active editor and publisher of psychology books and periodicals. Soon plans were laid to invite influential psychologists to contribute their life stories to a series of edited volumes entitled *A History of Psychology in Autobiography*. The rationale for the project was stated simply and directly: "Since a science separated from its history lacks direction and promises a future of uncertain importance, it is a matter of consequence to those who wish to understand psychology for those individuals who have greatly influenced contemporary psychology to put into print as much of their personal histories as bears on their professional careers."[1] The first volume, comprised of contributions from fifteen psychologists, was published in 1930. Two more volumes appeared, in 1932 and 1936, before the series was abandoned.[2]

In 1952, not long after the publication of the second edition of *A History of Experimental Psychology,* the series was revived.[3] Now Boring had his turn as a contributor, and he produced a remarkably candid account of his personal and professional life. He traced his career chronologically and ended with an attempt to put himself in perspective by outlining what he saw as the major aspects of his personality.

1 Carl Murchison, ed., "Preface," in *A History of Psychology in Autobiography,* vol. 1 (Worcester, Mass.: Clark University Press, 1930), pp. ix–x, quote on p. ix.

2 Carl Murchison, ed., *A History of Psychology in Autobiography,* vol. 2 (Worcester, Mass.: Clark University Press, 1932); idem, *A History of Psychology in Autobiography,* vol. 3 (Worcester, Mass.: Clark University Press, 1936).

3 Edwin G. Boring, Herbert S. Langfeld, Heinz Werner, and Robert M. Yerkes, eds., *A History of Psychology in Autobiography,* vol. 4 (Worcester, Mass.: Clark University Press, 1952).

Boring admitted that he was a compulsive person, driven by an unconscious as well as conscious desire to succeed. The key to his personality, he thought, was the tension between his conflicting needs for "power and achievement [versus] approval and affection." Here was where he identified his wish to become the "commanding servant" of the various groups of which he was a member. By fulfilling his own egocentric goals in the context of collective endeavor, he could escape the persistent sense of insecurity that had plagued him since childhood.[4]

Boring went on to ponder the question of whether he had achieved his elusive goal of maturity and decided that he was closer to the goal though he had not quite yet reached it. It was clear that his psychoanalysis in the early 1930s had helped him to gain some insight, even though it had not been as successful at relieving his anxieties as he had hoped. Boring had tried psychoanalysis because he was, in his words, "insecure, unhappy, frustrated, afraid." In spite of his professional achievements, he did not feel successful. He turned to psychoanalysis for relief, even though he was doubtful about its scientific veracity, and was analyzed by Hanns Sachs, a member of Freud's inner circle who had emigrated to Boston.

Perhaps more therapeutic than the analysis itself was Boring's public airing of his experience. A few years later he discussed his therapy in an article entitled "Was This Analysis a Success?" that appeared in a special issue of the *Journal of Social and Abnormal Psychology* (1940). Boring's article appeared with several others as part of a symposium on "Psychoanalysis as Seen by Analyzed Psychologists." The symposium provided a forum for academic psychologists to come to grips with Freudian theory by discussing their individual experiences. Although the participants focused on questions of scientific validity, the essentially subjective nature of their evidence represented a departure from conventional standards of experimental proof. It also suggested a reflexive turn in psychologists' thinking about themselves and their work that was nowhere more clearly demonstrated than in Boring's transmutation of his idiosyncratic encounter into a matter of broad professional import.

Boring, in common with many career-minded people, had trouble separating the personal and professional aspects of his life. Like Titchener, his quest for authority, in science and in life, knew few bounds. But in contrast to his mentor, Boring was willing to put his private anxieties on public exhibit, and thereby share his suffering with others. This trait, however, did not become manifest until the late 1930s when Boring extended his role as psychology's gadfly in new directions by using his own life and work as a lens through which to view larger issues facing the profession.

The confessional mode suited Boring well. In his vast private correspondence he worried constantly about where the profession was going. Was it remaining true

---

4 Edwin G. Boring, in *A History of Psychology in Autobiography,* vol. 4 (Worcester, Mass.: Clark University Press, 1952), pp. 27–53, quote on p. 51.

to the scientific vision of its founders, or were demands for practical knowledge turning it too much in applied directions? In his published writings, both formal and informal, he took pains to express his point of view clearly and fully. How, for instance, should graduate education be conducted? Should students be selected on the basis of their academic promise, or should programs endeavor to train any interested individual to become a psychologist? In his organizational efforts, locally as well as nationally, he fussed endlessly over the details of management and administration. What was the proper relation between experimental and clinical psychology, at Harvard and other universities? What steps should the American Psychological Association take toward the certification and licensure of consulting psychologists? All in all, Boring became perhaps the chief spokesman for the psychology community, a role that he grew into during the 1940s as he developed an ability to transform his personal preoccupations into public issues that seemed to articulate the concerns shared by many of his colleagues.

After retiring from Harvard in 1957, Boring took the opportunity to recast his life story once again and produced a book entitled *Psychologist at Large*. It contained an updated version of his earlier autobiographical essay, some reprinted articles and editorials, a few selections from his voluminous correspondence, and a bibliography of his writings.[5] Here was a portrait of a psychologist's life that refracted the history of the field. Its author's implied claim that he had become America's psychologist-at-large could hardly be disputed.

5 Edwin G. Boring, *Psychologist at Large* (New York: Basic Books, 1961).

10

The Mirror of Practice
Toward a Reflexive Science

In September 1939, just after the war in Europe broke out, Gordon Allport delivered his presidential address at the annual meeting of the American Psychological Association, held at Stanford and Berkeley, California. Reflecting on the historical symbolism of meeting for the first time on the shores of the Pacific, he noted that events across the Atlantic had placed "the burden of scientific progress in psychology" on the profession. Faced "with the responsibility for the preservation and eventual rehabilitation of world psychology," the Harvard professor asked, "Are we American psychologists equipped for the versatile leadership demanded by our comprehensive discipline?" He proposed to answer his rhetorical question through an analysis of "the psychologist's frame of reference" as it was reflected in the pages of the psychological literature over the preceding fifty years.[1]

Among the trends he observed was a rapid rise in the use of statistics, to the point that nearly half of the published literature relied on them. There was also a noteworthy increase in the use of animals as subjects and a concomitant growth in the proportion of methodological studies of all kinds. His data suggested "the development of a notable schism between the psychology constructed in a laboratory and the psychology constructed on the field of life."[2] In a footnote, Allport went on to call attention to the unwitting hypocrisy of some academic purists when they criticized applied psychology: "Outside the laboratory he lives a cultured and

---

1 Gordon W. Allport, "The Psychologist's Frame of Reference," *Psychological Bulletin*, 1940, *37*:1–28; reprinted in Ernest R. Hilgard, ed., *American Psychology in Historical Perspective: Addresses of the Presidents of the American Psychological Association, 1892–1977* (Washington, D.C.: American Psychological Association, 1978), pp. 371–395, quote on p. 372. The data were reported in greater detail in J. S. Bruner and G. W. Allport, "Fifty Years of Change in American Psychology," *Psychological Bulletin*, 1940, *37*:757–776.

2 Hilgard, *American Psychology in Historical Perspective*, pp. 383, 390n11.

varied life of a free agent and useful citizen. Yet his methodological work in the laboratory overspreads very little of his daily experience and prevents integration in his life. Though he generally repudiates a dualism of mind and body, he welcomes the equally stultifying dualism of laboratory and life." Thus Allport urged his fellow psychologists to consider the social ends as well as the scientific means of their work and to become more conscious of their behavior as scientists.

Concerned with what he would later term "methodolatry," Allport saw danger in the increasing emphasis on experimentation with infrahuman subjects, especially rats, which allowed for the use of precise, objective, and quantifiable techniques. In their rush to emulate methods identified with the natural sciences and their associated prestige, psychologists might lose sight of the real power of the older disciplines in *"predicting, understanding,* and *controlling* the course of nature for mankind's own benefit."[3] Allport argued that psychologists had hardly improved upon plain common sense in being able to usefully predict behavior. What was needed was more attention to the specifics of single cases and the contexts in which they are embedded, so that psychologists could "tell what will happen to *this* child's IQ if we change his environment in a certain way, whether *this* man will make a good executive, whither *this* social change is tending."[4] In a similar fashion, he argued that psychologists had hindered their understanding by not considering sufficiently the subject's point of view or frame of reference, and by interpreting their research findings within narrow methodological constraints. This scientific tunnel vision enabled psychologists to rationalize their lack of concern with the wider implications of their work.

In the realm of the control of behavior, applied psychologists had been most active, but it was unclear how much of their success rate could be attributed to the adequacy of their scientific skills. And outside of areas such as clinical psychology, much of what was published in the psychological literature had little apparent relevance to the problems faced by ordinary people. Rather than retreating from such difficulties, Allport urged renewed concentration on the practical control of human affairs, which ultimately justified the pursuit of psychological science.[5]

Allport was a key figure in the expansion and legitimization of "sociotropic" psychology before the war. In particular, he championed the idiographic approach (i.e., focus on individual cases) to psychology to augment the prevailing nomothetic orientation (i.e., the search for general principles). In his 1937 textbook on the psychology of personality and in his 1939 APA presidential address, he argued

---

3 Ibid., p. 385 (emphasis in original).

4 Ibid., p. 386 (emphasis in original).

5 Ibid., pp. 390–391. Another effort to characterize American psychology immediately prior to World War II found it to be "increasingly empirical, mechanistic, quantitative, nomothetic, analytic, and operational." See A. G. Bills, "Changing Views of Psychology as a Science," *Psychological Review,* 1938, *45*:377–394.

that psychology must take into account the individual as a unit of analysis and explore the rich complexity of case studies on their own terms. He believed that behavioral principles could be gleaned from the study of single cases without recourse to statistical aggregation.[6]

In 1942 Allport developed his ideas about idiographic psychology further in a technical monograph, "The Use of Personal Documents in Psychological Science." Embedded in his discussion of various forms of personal documents, such as autobiographies, diaries, letters, and the like, Allport made a case for incorporating personality into psychology, both on the basis of its scientific fruitfulness and its practical implications. He insisted that psychologists should stay in close touch with the "concrete" psychology encountered in daily life; only it provided a proper basis for accurate and useful knowledge. Furthermore, the idiographic approach brought "psychological science closer in line with the ethics of democracy," with its stress on the individual.[7]

The Triumph of Statistics

Support for Allport's thesis about the increasing influence of statistical methods in psychology was provided by a review article published in the *American Journal of Psychology* in 1938 by Jack W. Dunlap, secretary of the newly formed Psychometric Society. As such techniques proliferated, they placed new demands on psychologists to become more mathematically sophisticated. Although fully committed to the incorporation of statistics into psychology, Dunlap was aware of the burdens it might entail. He commented ironically, "I can only extend my sympathy to the psychologist of the future, for it seems as if he must first be a mathematician, then a statistician, . . . and, if he is not dead of old age by then, a psychologist."[8]

Dunlap's projection was partly on target. Statistical methodology did become increasingly significant in psychology. What began as an important movement in psychometrics spread more generally through experimental psychology in the 1940s. Its influence on research design and analysis can be gauged by the fact that by the early 1950s coursework in the analysis of variance had become a standard feature of graduate training in the United States.[9] Its routine use, standardized and

6 Gordon W. Allport, *Personality: A Psychological Interpretation* (New York: Henry Holt, 1937), pp. 19–23.
7 Gordon W. Allport, *The Use of Personal Documents in Psychological Science* (New York: Social Science Research Council, 1942); quote on p. 148.
8 Jack W. Dunlap, "Recent Advances in Statistical Theory and Applications," *American Journal of Psychology,* 1938, *51:*558–571, quote on p. 57.
9 Anthony J. Rucci and Ryan D. Tweney, "Analysis of Variance and the 'Second Discipline' of Scientific Psychology: A Historical Account," *Psychological Bulletin,* 1980, *87:* 166–184, on pp. 179–180.

codified in textbooks, also meant that psychologists did not necessarily have to understand the complexities of statistics in order to apply them to their work. Indeed, textbooks on psychological statistics contained a curious hybrid of the approaches of Fisher and of Neyman and Pearson, each of which reflected different assumptions.[10] These conceptual incompatibilities were ignored in the interest of providing a tidy "cookbook" approach to the subject.

As psychologists attempted to deal quantitatively with individual behavior, they were faced with enormous problems of inconsistency. People behaved differently at different times, and what they did in one experimental setting might have little relation to what they did in another. Statistical treatment of group data offered a way out. By combining data from many individuals, certain patterns might emerge as characteristic of the statistical group, even though the actual behavior of any particular member of the group might not conform to the statistical "average." As Kurt Danziger has pointed out, by the 1930s the "triumph of the aggregate" extended through the entire range of empirical research practices in psychology. It served to reinforce the epistemological autonomy of the discipline. In Danziger's words, "the artificial constitution of collectivities *by the research process itself* provided the basis for a science of psychological abstractions that need never be considered in the context of any actual individual personality or social group." It was an essential component in the construction of the collective subjects upon which psychology's claims to knowledge were founded.[11]

After World War II the trend toward group data became overwhelming. Between 1940 and 1960, the articles published in both the *Journal of Experimental Psychology* and the *Journal of Clinical Psychology* showed a dramatic increase in the reporting of data attributable to groups rather than to individuals. By 1960 only about 5 percent of the empirical studies in either journal reported any individual data at all.[12]

The rise in the reporting of group data coincided with the adoption of the analysis of variance (ANOVA) statistical techniques for handling multiple-variable research designs. Although researchers had explored the use of multifactor statistical tests in psychophysics, reaction time studies, and other areas, it was not until the 1940s that a concerted effort was made to construct a conceptual rationale for their use. In 1940 Richard Crutchfield and Edward Tolman of the University of California published a short paper citing the advantages of a multifactor approach to research design that echoed R. A. Fisher's arguments in *The Design of Experiments*

10 Gerd Gigerenzer and David J. Murray, *Cognition as Intuitive Statistics* (Hillsdale, N.J.: Lawrence Erlbaum Associates, 1987), chap. 1.

11 Kurt Danziger, *Constructing the Subject: Historical Origins of Psychological Research* (New York: Cambridge University Press, 1990), p. 85 (emphasis added).

12 James H. Capshew, "Constructing Subjects, Reconstructing Psychology," *Theory & Psychology,* 1992, 2:243–247.

(1935). They went beyond Fisher, however, in suggesting that ANOVA could be used for modeling complex behavioral systems. In particular, they cited Tolman's theory, which relied on hypothetical "intervening variables" that mediated between observable independent and dependent variables. Thus they argued for "the unique significance of multiple-variable designs in the study of those areas of behavior where it is known or suspected that complex interaction of variables exists."[13]

A review essay published in 1943 counted more than forty studies that utilized ANOVA designs.[14] With the spread and adoption of statistical techniques for the analysis of variance, the psychology community achieved a measure of methodological consensus soon after the war. In their search for scientific certainty, psychologists had moved from attempting to define their work in terms of content (whether in terms of consciousness or behavior) to an almost obsessive concern with research design and quantitative methods. The rise of the analysis of variance represented another step along the path of methodological abstraction and the creation of a universal tool for psychological investigations. In short, as psychologists fervently embraced statistics for its methodological charms, "the dream of the scientist who arrives at new knowledge by a completely mechanized process seemed to have become real."[15]

Clinical Versus Statistical Prediction

In their efforts to legitimate themselves as a mental health profession, clinical psychologists naturally stressed their scientific outlook and training. In particular, they claimed that their commitment to *scientific* problem solving provided the basis for their unique contribution as a helping profession. But clinical psychology, in theory and in practice, raised important methodological issues.

Correlational methods utilizing inferential statistics were already well entrenched in the psychometric tests commonly used in psychological diagnosis. Clinical experience, however, was also considered an important factor, both in psychodiagnosis and psychotherapy. Some clinicians, such as Carl Rogers, focused almost

---

13 R. S. Crutchfield and E. C. Tolman, "Multiple-variable Design for Experiments Involving Interaction of Behavior," *Psychological Review,* 1940, *47:*38–42. For a detailed survey of the literature during this period, see A. D. Lovie, "The Analysis of Variance in Experimental Psychology: 1934–1945," *British Journal of Mathematical and Statistical Psychology,* 1979, *32:*151–178.

14 H. E. Garrett and J. Zubin, "The Analysis of Variance in Psychological Research," *Psychological Bulletin,* 1943, *40:*233–267. The rapid growth in the use of ANOVA documented by this paper led Lovie, "Analysis of Variance," p. 157, to state: "It seems that by 1943 the battle over the widespread acceptance of ANOVA had been fought and won."

15 Gigerenzer and Murray, *Cognition as Intuitive Statistics,* p. 27. Although the authors identify and analyze the structure of the "inference revolution" between 1940 and 1955, they do not discuss how it occurred.

exclusively on the development of clinical judgment, eschewing diagnostic categorization based on standardized testing.

Among the most acute students of such issues was Paul Meehl. Trained entirely at the University of Minnesota (Ph.D. 1945) and employed as a faculty member since, he had a strong background in both experimental and clinical psychology. An accomplished scientist-practitioner with a strong interest in the philosophy of science, Meehl staked out psychodiagnosis as his special territory. He contributed to the elaboration of the Minnesota Multiphasic Personality Inventory (MMPI), which came into widespread use following the war, and began to examine critically psychodiagnostic concepts and methods.[16]

In 1954 Meehl published a monograph, *Clinical versus Statistical Prediction: A Theoretical Analysis and a Review of the Evidence.* In it he addressed the fundamental conflict he had encountered as a scientist-practitioner between the immediate demands for useful knowledge confronted in clinical practice and the equally insistent concern over scientific adequacy.[17] Although he was careful to present a balanced treatment of the strengths and weaknesses of both approaches, many clinicians took the work as a rejection of the creative and imaginative role of the clinician in favor of actuarial methods. For his part, Meehl was not interested in "proving" that a single approach was better, but rather to determine how to increase accuracy and efficiency in diagnostic prediction.

In a later paper Meehl contrasted the "rule-of-thumb" method typical of psychodiagnosis, in which a clinician reads and integrates the result of various personality tests, with a "cookbook" approach that assigned numerical values to psychometric data based on explicit rules of association. In advocating the development of actuarial tables and diagnostic cookbooks, Meehl displayed his penchant for the use of mathematics as a useful tool in psychology. There was also work to be done on quantifying the inductive probability of clinical judgments. But above all, Meehl was pragmatic: "Shall we use our heads, or shall we follow the formula? Mostly we will use our heads, because there just isn't any formula. . . ."[18] In his quest for better formulas, Meehl expressed his concern over the methodological foundations of clinical practice as the ultimate basis for professional status.

Toward a Science of the Person

As the technoscientific project of postwar psychology evolved, the symbiotic relationship between its scientific and its professional components inevitably

---

16 See Roderick D. Buchanan, "The MMPI and Personality Assessment in American Clinical Psychology, 1935–1965" (Ph.D. diss., University of Melbourne, 1992).

17 Paul E. Meehl, *Clinical versus Statistical Prediction: A Theoretical Analysis and a Review of the Evidence* (Minneapolis: University of Minnesota Press, 1954).

18 Paul E. Meehl, *Psychodiagnosis: Selected Papers* (New York: Norton, 1973), p. 89.

experienced stresses and strains. Psychologists used various means to manage such problems and to harness the energy that their wartime success had generated. They created conceptual schemes to handle its complexity and established organizational structures to channel its power.

The identity of psychology, in theory as well as in practice, was constantly at issue. In particular, the epistemological values embodied in psychological research were undergoing redefinition as the professional activities of psychologists became more explicitly motivated by applied interests. There was a general trend toward defining psychology as the science of the person, of conceptualizing individuals as biosocial entities. Furthermore, the framework of reductionistic behaviorism was loosening to allow for consideration of unobservable mental events – at least in theoretical formulations if not in method – which led to the rise of cognitive psychology.[19]

In this environment several interesting efforts were made to nurture the symbiosis between research and practice by reconfiguring the relations among psychologists, their science, and their subjects. Psychologists became increasingly aware of the reflexive nature of their science, that their behavior as psychologists was a factor to be considered in the laboratory and in the clinic. After all, if psychology had any claim to a general science of behavior, then it must also encompass the behavior of psychologists as scientists and practitioners. Psychologists across a broad spectrum sought to exploit this unique feature of their science and integrated it into their work in a variety of ways.

The challenges of reflexivity had been recognized before World War II, but then it had been generally viewed as a potential problem, to be managed through methods and procedures that would minimize the "subjective" aspects of psychological investigation and its resultant knowledge. The radical epistemological implications of a deeply reflexive psychological science were largely avoided.[20] After the war, as the social esteem and self-confidence of psychologists soared, reflexivity was embraced rather than shunned as an integral property of scientific psychology. Although this embrace took many forms, its many varieties were based on a positive valuation of its potential.

Such developments were due, in part, to the incorporation of psychoanalytic theory into academic psychology. Before World War II, psychoanalytic practice had been almost the exclusive province of the medical profession. Ideas derived from Freud were shunned by many academic psychologists as not being sufficiently scientific, despite widespread popular interest in American society. During the 1930s,

---

19 See Howard Gardner, *The Mind's New Science: A History of the Cognitive Revolution* (New York: Basic Books, 1985); Ernest R. Hilgard, *Psychology in America: A Historical Survey* (San Diego, Calif.: Harcourt Brace Jovanovich, 1987), chap. 7.

20 Jill G. Morawski, "Self-Regard and Other-Regard: Reflexive Practices in American Psychology, 1890–1940," *Science in Context,* 1992, 5:281–308.

however, the situation began to change. An increasing number of American intellectuals, including psychologists, were psychoanalyzed. And European intellectuals, fleeing from political repression, found their way to refuge in the United States and brought with them a greater appreciation for the psychoanalytic point of view. This two-way traffic paved the way for a rapprochement between psychoanalysis and academic psychology.[21]

In the context of World War II the links became even closer. In 1940 several psychologists shared their experiences with psychoanalysis in a special issue of the *Journal of Abnormal and Social Psychology.*[22] Ostensibly, the purpose was to evaluate the scientific and therapeutic value of the system; it was also a sign of acceptance. Psychoanalysis proved itself on the battlefield as well, as psychiatrists found it a useful form of therapy. After the war, these developments contributed to the growing impact of psychoanalysis on academic psychology.

As postwar expansion buoyed their hopes for the future, American psychologists, already prone to self-reflection, turned their attention inward in new ways. For instance, studies of the psychology of the scientist (including the psychologist) and research into the social psychology of experimentation (including psychological research) were made. The American Psychological Association sponsored a major self-study of the scientific and professional dimensions of psychology in an attempt to shed light on the tremendous growth of psychology.

Individually and collectively, psychologists expressed their concern with epistemological questions raised by their work by holding the mirror of science up to themselves. They assumed that if scientific behavior was simply one among many patterns of human behavior, then it should be possible to construct a general psychology of the scientist that would apply to psychologists as well as to other kinds of scientists. As the self-consciousness of psychologists about themselves and their work increased, their reflections took a more thoroughly reflexive turn. That is, there was a greater appreciation that self-analysis itself was not a simple and

21 See John C. Burnham, "The Influence of Psychoanalysis upon American Culture," in Jacques M. Quen and Eric T. Carlson, eds., *American Psychoanalysis: Origins and Development* (New York: Brunner/Mazel, 1978), pp. 52–72; reprinted in Burnham, *Paths into American Culture: Psychology, Medicine, and Morals* (Philadelphia: Temple University Press, 1988), pp. 96–112; Nathan G. Hale, Jr., *The Rise and Crisis of Psychoanalysis in the United States* (New York: Oxford University Press, 1995). American psychologists who were psychoanalyzed included: Margaret Brennan, Edwin Boring, Sybil Escalona, Elsa Frenkel-Brunswick, Robert R. Holt, Carney Landis, Robert Lindner, Lester Luborsky, Neal E. Miller, O. H. Mowrer, Werner Muensterberg, Henry Murray, David Rapaport, Roy Schafer, Robert Sears, David Shakow, Donald P. Spence, and Milton Wexler; personal communications, Ernest R. Hilgard, 22 September 1993; Nathan G. Hale, Jr., 22 April 1994.

22 "Symposium: Psychoanalysis as Seen by Analyzed Psychologists," *Journal of Abnormal and Social Psychology,* 1940, *35.*

straightforward application of psychological techniques, but that the process of objectifying themselves and their work itself had some interesting consequences. For psychologists to use their own lives and experiences as sources of data for exemplifying their theories was one thing; to warrant their claims to knowledge by appealing to personal experience was another.

Clinical Studies of the Psychology of the Scientist

It was perhaps inevitable that the idiographic approach to psychological knowledge, defined by an interest in describing the specifics of an individual case rather than discovering general (i.e., nomothetic) principles, would be applied by psychologists to themselves. The rise of personality theory and tests since the 1930s, coupled with a long-standing tradition of autobiographical reflection by prominent psychologists, provided the background to the emergence of psychological studies of psychologists after the war.[23]

Such studies were part of a more general interest in exploring the personality and motivations of scientists that had emerged in the wake of World War II. One of the first researchers in the area was clinical psychologist Anne Roe (1904–1981). A few years after earning her doctorate from Columbia in 1932, she married the paleontologist George Gaylord Simpson and did not pursue regular academic employment. Instead she conducted a series of independent research projects, most notably a study of the effects of alcohol consumption upon eminent painters. Her broader interests lay in the relation between vocational choice and personality structure.

After the war Roe turned her attention to scientists, a group to which she had many close connections, through her husband's as well as her own associations. Noting that "what kind of a person the scientist is and why and how he becomes a scientist had never been seriously studied," she drew up plans to explore the topic

---

23 Another, perhaps less obvious, expression of interest in reflexivity among psychologists can be seen in the series of volumes of *A History of Psychology in Autobiography*. Launched in 1930, three volumes containing the personal reflections of over forty prominent psychologists had been published by 1936. After a hiatus of several years, the series resumed in 1952, reaching a total of eight volumes by 1989. The title of the series suggests the conceptual rationale – that the story of scientific psychology can best be told by those who have contributed most to its development, and that the microcosm of an individual life can illuminate the macrocosm of the discipline. Thus the natural interest in the people behind the research could be elevated by this appeal to a larger purpose. Whereas autobiography as a genre may exhibit only a primitive form of reflexivity – in the form of personal and disciplinary self-analysis – for psychologists it held a special place, because any account of a life history has to depend on some theory of human nature, even if implicit or unacknowledged.

using various clinical techniques. She received a major grant from the U.S. Public Health Service and launched a study of sixty-four eminent scientists.[24]

Her pool of subjects was selected with the aid of other leading scientists and was limited to those in fields of basic rather than applied research. The entire group was male and was subdivided into physical scientists, biologists, and social scientists (i.e., psychologists and anthropologists). The psychologists represented areas of experimental psychology and were chosen with the assistance of Edwin Boring, Ernest Hilgard, Lewis Terman, David Shakow, Jean MacFarlane, and Donald Lindsley.[25] Her research included extensive interviews with each scientist, including discussion of personal history and family background as well as scientific work habits. Roe utilized two projective tests, the Rorschach inkblot and the Thematic Apperception Test, as a means to probe personality dynamics. In order to gain information on intelligence, Roe administered a special Verbal-Spatial-Mathematical test devised with the help of the Educational Testing Service.

Roe presented the results of her research in the book *The Making of a Scientist,* published in 1952. Weaving together demographic, interview, and test data, she described similarities and differences among her three subgroups. Written in the form of a narrative, the book uses vignettes from individual subjects to illustrate general themes and conclusions about the psychology of scientists.

Although she felt more at home with the psychologists and anthropologists, Roe treated the social scientists much as she did the physicists and biologists. Because she selected basic research scientists, none of the psychologists in her sample were working in social or clinical fields. She did note that, generally speaking, the psychologists were more attuned to the interpretive issues raised by her study. Substantively, Roe found that the social scientists were more interested in people than things, that the physicists tended to think symbolically, and that the biologists were oriented toward rational control and issues of form.

Apparently Roe was not very self-conscious about the reflexive implications of her research; it seemed to be simply an application of known techniques to a new population. In her concluding chapter, "What Does It Mean for You?" she attempted to draw lessons for the general public. Celebrating the freedom of inquiry enjoyed by scientists and the rewards of research, Roe argued that such satisfactions were open to everyone. It was not necessary to become a scientist in order "to develop scientific attitudes, which will make you a better and a happier citizen. Research in the broadest sense is more a habit of mind and a method of approach to problems than a specific technique. Certainly there is nothing esoteric about it

---

24 Anne Roe, *The Making of a Scientist* (New York: Dodd, Mead, 1952), p. 12.
25 Ibid., p. 51. The psychologists selected were Gordon Allport, Frank Beach, Jerome Bruner, Clarence Graham, J. P. Guilford, Harry Harlow, Ernest Hilgard, Karl Lashley, Donald Lindsley, Curt Richter, Carl Rogers, Robert Sears, B. F. Skinner, and S. S. Stevens; Anne Roe Papers inventory, American Philosophical Society Library, Philadelphia, Pa.

(as I hope this book has demonstrated about clinical psychological research, at least)."[26] Thus Roe attempted to assimilate science to common sense and in the process redrew the boundaries of psychology to include what might be considered a relatively primitive form of disciplinary self-analysis.

A few years later another clinical psychologist undertook a similar study of the psychology of research scientists. Bernice Eiduson, who also happened to be married to a scientist, sought "to strip rationalization from reality" and replace stereotyped views of scientists with empirically derived knowledge of their personalities, thinking styles, and self-images.[27]

Eiduson used research methods similar to Roe's but took a different approach. Her sample of forty research scientists was deliberately heterogeneous in terms of their disciplinary affiliation, professional reputation, and academic status. They came from physics, geoscience, chemistry, and the biological sciences, and were all affiliated with West Coast academic institutions. Her focus was on the psychological factors that led to their choice of a research career. She did not use a criterion of eminence like Roe, although many of her subjects were successful according to conventional standards of achievement.

Like Roe, Eiduson was interested in making generalizations about generic psychological characteristics while maintaining her focus on the particularities of each individual case. She found intellectual giftedness and an impersonal emotional climate to be common developmental features in the group. In terms of cognitive style, she concluded that the scientists were "intellectual rebels," questioning received opinion, on the lookout for novelty, always seeking new ways to test and describe their ideas. In terms of their life-style, most outside activities revolved around maximizing scientific productivity. Eiduson identified the characteristically narrow focus of scientific workers, concluding that "researchers have, in general, restricted their roles, duties, and responsibilities to their own fields; thus fixating on becoming better specialists, more competitive, more imaginative, but only in their own areas of expertness."[28]

The studies of both Roe and Eiduson can be seen as part of a broad concern about "scientific manpower" that had arisen because of World War II. Problems of supply and demand, recruitment and retention, education and employment, and the like were augmented by interest in the inner life of scientists – their motivations, attitudes, and cognitive structures – in short, their psychology. Indeed, Eiduson was explicit in her contention that, as the external rewards for a scientific career approximated those of other professions, it would be necessary to focus on the internal

26 Ibid., p. 243.
27 Bernice T. Eiduson, *Scientists: Their Psychological World* (New York: Basic Books, 1962), p. 3.
28 Ibid., p. 258.

rewards afforded by scientific research and try to foster interest among students on that basis.[29] These studies also indicated the feasibility of a psychology of science, although such a field has been slow to develop.[30] The reflexive connotations of such work have not been widely recognized in psychology; other investigative practices revealed the challenges and opportunities of reflexivity more quickly and clearly.

Experimenter Effects

Another expression of the general movement toward greater reflexivity was the identification of experimenter effects in psychological research. The psychological experiment itself became a site for investigation as researchers discovered that the expectations of the researcher as well as those of the subject could affect the outcome of experiments with human beings.

The suggestion that the experimental situation itself be considered a psychological problem was made by Saul Rosenzweig in an article published in the *Psychological Review* in 1933. The young Harvard Ph.D. raised the possibility that biased attitudes, conscious or not, on the part of the experimenter or the subject might influence the outcome of the experiment – a possibility that other researchers had previously identified. He pointed out the special challenge of psychological research: "when one works with human materials one must reckon with the fact that everyone is a psychologist." Rosenzweig went on to outline ways to explore such influences systematically, but he did not pursue the issue further, nor did his article generate much response among other psychologists.[31] Sustained research on the topic did not occur until two decades later.

29 Ibid., p. 265.
30 See W. R. Shadish and S. Fuller, eds., *The Social Psychology of Science* (New York: Guilford, 1994), on the current state of the art.
31 Saul Rosenzweig, "The Experimental Situation as a Psychological Problem," *Psychological Review,* 1933, *40*:337–354, quote on p. 342. The Hawthorne experiments, conducted by Elton Mayo and his Harvard associates around this time, raised similar questions. One of the roots of postwar interest in reflexivity can be traced to the discovery of the "Hawthorne effect" in the 1930s. The notion that the very process of studying people could affect their behavior in significant ways had been popularized by the team of industrial sociologists who had discovered the "Hawthorne effect" in their research on worker productivity. Completed in the late 1930s, the study of Bell Telephone's Hawthorne plant near Chicago became widely known following World War II and provided support for the emerging management ideology known as "human relations." In brief, the Hawthorne researchers discovered that, in the process of systematically investigating different variables affecting the work situation on the shop floor, productivity increased due to the attention paid to the workers by the researchers. See Richard Gillespie, *Manufacturing Knowledge: A History of the Hawthorne Experiments* (Cambridge: Cambridge University Press, 1991).

One major line of research grew out of Robert Rosenthal's dissertation research in the mid-1950s. As a UCLA graduate student Rosenthal was inspired by the work of Sigmund Freud and Henry Murray to develop an experimental test of the defense mechanism of projection. He noticed that his knowledge of which treatment group each subject was assigned to somehow affected their subsequent performance on the experimental task. When he shared this finding with his advisors, they dismissed it as a methodological annoyance to be ignored. The phenomenon had been noted earlier by investigators such as Ebbinghaus and Pavlov, but it had never been systematically investigated. So Rosenthal took up the issue of "unconscious experimenter bias" and conducted a series of research studies that explored how expectations on the part of the experimenter affected the behavior of experimental subjects. On the basis of such demonstrated effects in the laboratory, Rosenthal suggested that similar influences might be at work among doctors, psychotherapists, teachers, and other practitioners.[32]

Another line of work focused more on the impact of attitudes and expectations of the subject in the experimental situation. In experiments on hypnosis that he conducted in the 1950s, Martin Orne found that subjects tended to give experimenters the responses they thought the experimenters wanted, on the basis of subtle cues and hints. Such "demand characteristics" of the experimental situation were explored for their intrinsic interest as well to devise ways to control their confounding effects.[33]

In general, such work was aimed toward eliminating or minimizing such experimental "artifacts" through refinements in method rather than toward consideration of their epistemological implications. By the late 1960s a network of researchers interested in the social psychology of the experiment had emerged. Their goal was to understand and thereby minimize the "artifact threat" posed by the experimenter–subject relationship. Again, the more general reflexive implications of the work were not emphasized.[34]

32 Robert Rosenthal, "On the Social Psychology of the Psychological Experiment: The Experimenter's Hypothesis as Unintended Determinant of Experimental Results," *American Scientist,* 1963, *51:*268–283; idem, "On Being One's Own Case Study: Experimenter Effects in Behavioral Research – 30 Years Later," in Shadish and Fuller, eds., *The Social Psychology of Science,* pp. 214–229.

33 M. T. Orne, "On the Social Psychology of the Psychological Experiment: With Particular Reference to Demand Characteristics and Their Implications," *American Psychologist,* 1962, *17:*776–783; idem, "Hypnosis, Motivation, and the Ecological Validity of the Psychological Experiment," in William J. Arnold and Monte M. Page, eds., *Nebraska Symposium on Motivation* (Lincoln: University of Nebraska Press, 1970), pp. 187–266.

34 R. Rosenthal and R. L. Rosnow, eds., *Artifact in Behavioral Research* (New York: Academic Press, 1969). See Danziger, *Constructing the Subject,* pp. 59–64, 174–175. See also Jerry M. Suls and Ralph L. Rosnow, "Concerns about Artifacts in Psychological Experiments," in Jill G. Morawski, ed., *The Rise of Experimentation in American Psychology* (New Haven, Conn.: Yale University Press, 1988), pp. 163–187.

Group Dynamics

If the circumstances of experimental research provided the basis for a limited exploration of reflexivity in psychology, then applied work was where such issues could not be easily ignored. Dealing with individual persons, often on a one-to-one basis, heightened psychologists' awareness of the subjective dimensions of behavior – their own as well as that of their clients. It also underscored the essentially relational nature of behavior change in general, and in psychotherapy in particular. With the growth of applied psychology, especially clinical, it was perhaps inevitable that widespread changes in professional practices would have significant epistemological consequences.

As Kurt Lewin had argued toward the end of the war, the progress of psychology depended upon a close interaction between theory and practice. After the war, he and his co-workers tried to pursue this goal at the Research Center for Group Dynamics through various forms of "action research." One early project involved training community leaders to combat religious and racial prejudice. In the summer of 1946, Lewin, who also was the guiding spirit behind the Commission on Community Relations, the research arm of the American Jewish Congress, supervised a leadership workshop for the Connecticut State Inter-Racial Commission. The forty-one participants, about half of whom were African American or Jewish, were drawn mainly from the ranks of educators and social services workers, with a few others from business and labor. Members of the group "hoped to develop greater skill in dealing with other people, more reliable methods of changing people's attitudes, insight into reasons for resisting change, a more scientific understanding of the causes of prejudice, and a more reliable insight into their own attitudes and values."[35]

Lewin and his colleagues considered the workshop a large-scale "change" experiment in which the process of training could be monitored and evaluated on an ongoing basis to provide research data on group dynamics. At the end of each daily session, the training and research staff members would review and discuss their observations of the trainees' behavior. In the course of one of these post-session conferences, a few of the trainees happened to be present. As they shared with the assembled psychologists their perceptions of what had transpired during the day, it was clear that they interpreted things differently than the experts. Excited by this new information, Lewin and his colleagues continued the discourse with the participants and came to the conclusion that the very process of analyzing leadership training sessions with members of the group could have a therapeutic function. This was an example of how social psychological investigation could, through the process of mutual feedback, lead to desirable changes both in the personalities of

---

35 Alfred J. Marrow, *The Practical Theorist: The Life and Work of Kurt Lewin* (New York: Basic Books, 1969), p. 211.

individual participants and in their interpersonal relations within the group. Thus the study and the enhancement of group dynamics went hand in hand.[36]

This initial experiment was so successful that the approach was institutionalized in the National Training Laboratories, established in 1947 in Bethel, Maine, with an initial grant from the Office of Naval Research. Here T (for training) group research and practice continued, contributing to the proliferation and spread of "sensitivity training" in various forms. By the late 1960s Carl Rogers was calling the technique "the most significant social invention of this century."[37]

In exploring the phenomena of group dynamics, the Lewinians discovered that research itself was a form of human relations. Moreover, it was a peculiarly powerful means to promote behavioral change. By making the group process itself the focus of the group, the hermeneutic circle was completed, with group dynamics feeding off of itself. The tension between participation and observation (also encountered in anthropological fieldwork) could be creatively resolved by mutual feedback among all parties involved.

Lewin's work demonstrated the close connections existing between social psychology and the psychology of personality as researchers sought to explain individual behavior in terms of social context. Although the group was the focus for such work, it was the individual that was changed. In psychotherapy, the person was often the explicit object of interest, and psychologists tended to formulate their ideas about personality and therapy in terms of individual psychodynamics.

Carl Rogers and Client-Centered Therapy

Before the war, Henry Murray at the Harvard Psychological Clinic had advocated the development of "personology," or the systematic study of personality.[38] He was not alone, of course, in bringing the psychology of personality to new prominence, and in the postwar context it flourished. Among the most influential attempts to construct a psychology of the person was that of Carl Rogers.

Rogers spent the years during World War II on the faculty of Ohio State University. He arrived at the university in 1940 as a full professor of psychology, after nearly a dozen years working as a counselor in Rochester, New York, first for the Society for the Prevention of Cruelty to Children and then as the founding director

36 Ibid., pp. 210–214.
37 See Leland P. Bradford, Jack R. Gibb, and Kenneth D. Benne, eds., *T-Group Theory and Laboratory Method* (New York: Wiley, 1964); Kurt W. Back, *Beyond Words: The Story of Sensitivity Training and the Encounter Movement* (Baltimore: Pelican, 1973; orig. pub. 1972); Carl R. Rogers, "Interpersonal Relationships U.S.A. 2000," *Journal of Applied Behavioral Science,* 1968, *4:*208–269; quoted in Marrow, *Practical Theorist,* pp. 213–214.
38 See Edwin S. Schneidman, *Endeavors in Psychology: Selections from the Personology of Henry A. Murray* (New York: Harper & Row, 1981).

of the Rochester Guidance Center. Trained in the eclectic environment of Teachers College, he received his Ph.D. in 1931 from Columbia. By the late 1930s he was beginning to articulate his nondirective or client-centered approach to psychotherapy. He was among the pioneers in establishing the specialty of clinical psychology.

In 1942 Rogers published *Counseling and Psychotherapy,* contrasting his new nondirective approach with more traditional directive approaches. He took an explicitly scientific stance in presenting his work as a "series of hypotheses in regard to counseling which may be tested and explored."[39] One of his fundamental assumptions was that the "client" (in contrast to the traditional "patient") was the best guide to the identification and solution of his or her own problems. In this view, the counselor's role was to help establish a therapeutic relationship with the client. Rogers identified four basic ways in which counselors should try to create such relationships: genuine acceptance of the client as a person; a nonjudgmental attitude; the maintenance of limits in therapy; the absence of coercion or pressure.[40] The author's empirical bent was revealed by his inclusion of a complete transcript, totaling 176 pages, of a complete series of therapy sessions with a single client. The idea was that such a verbatim record of the responses of both the counselor and the client would best reveal the dynamics of the therapeutic process.

In 1944 Ralph W. Tyler, dean of the social sciences at the University of Chicago, invited Rogers to come to the university and organize a student counseling program. The next year Rogers moved to Chicago to establish the Counseling Center; its purpose was "to assist the student to help himself, to aid him in becoming more intelligently self-directing."[41] Although counseling was the order of the day, the staff was also committed to therapeutic innovation through research and organizational experimentation. The group of colleagues and graduate students gathered around Rogers enjoyed unusual esprit de corps, which was reinforced by the antagonism of the Department of Psychiatry. The relationship with the Department of Psychology was better, however. Chaired by James G. Miller, its members were actively engaged in the interdisciplinary exploration of cybernetics, general systems theory, and other postwar schemes that promised to unify the understanding and control of behavior.

In 1951 Rogers published his next major work, *Client-Centered Therapy: Its Current Practice, Implications, and Theory.* As its subtitle suggests, Rogers was concerned to ground his scientific generalizations in the firm territory of therapeutic experience. The book further developed the idea that the therapist was supposed to become an agent of psychotherapeutic change by embodying certain values and orientations, not simply by applying methods or techniques of psychotherapy. In

39 Carl R. Rogers, *Counseling and Psychotherapy* (Boston: Houghton Mifflin, 1942), p. 17.
40 Ibid., pp. 88–89.
41 Carl R. Rogers, in *The History of Psychology in Autobiography,* vol. 5 (New York: Appleton-Century-Crofts, 1967), pp. 343–384, quote on p. 363.

Rogers's words, the effective counselor "holds a coherent and developing set of attitudes deeply imbedded in his personal organization."[42]

Prompted by the competing demands of scientific research and professional practice, Rogers reported experiencing an ambivalence between "sensitively subjective understanding" and "detached objective curiosity" about people. In 1955 he expressed his concerns in the pages of the *American Psychologist* as a philosophical choice between "Persons or Science?"[43] Like the Lewinians, Rogers broke down the scientific/professional distinction by utilizing experiential data. Such data, however, were acquired through individual therapy rather than group interactions. For Rogers, it was necessary that the therapist possess certain personal qualities that would allow him or her to enter into a therapeutic relationship with a client. As the source of behavioral change was located within the client, and the therapeutic relationship was simply a means to facilitate positive change, the therapist served only as a catalyst, not as an expert who applied certain techniques. Although the client-centered approach seemed to downplay the professional role of the therapist, it had the effect of privileging the status of the scientist, who by virtue of his/her "experimental" approach to human relations (i.e., making hypotheses about the client and testing or checking them) represented a model of human behavior to emulate. Thus Rogers equated knowledge gained from experience with knowledge gained from experiment, as the therapist's self became an analogue for the research laboratory.[44]

Unable to resist the attraction of writing his own job description, Rogers moved to the University of Wisconsin in 1957. There he was involved in training both psychologists and psychiatrists, and launched an ambitious research program on the effects of psychotherapy on schizophrenics. It was during this period that he completed *On Becoming a Person* (1961), a mature statement of his views on personality development and the process of psychotherapy. Rogers expressed his conviction that "If I can provide a certain type of relationship, the other person will discover within himself the capacity to use that relationship for growth, and change and personal development will occur."[45]

In 1964 Rogers left university life and accepted a position at the Western Behavioral Sciences Institute, a nonprofit organization founded in 1959 by a Caltech physicist and others to explore issues in human relations. Here he became a leading figure in the humanistic psychology movement.

---

42 Carl R. Rogers, *Client-Centered Therapy: Its Current Practice, Implications, and Theory* (Boston: Houghton Mifflin, 1951), quoted on p. 19.

43 Rogers, *History of Psychology in Autobiography,* vol. 5, pp. 378–379; idem, "Persons or Science? A Philosophical Question," *American Psychologist,* 1955, *10:*267–278.

44 Carl Rogers, "Toward a Science of the Person," *Journal of Humanistic Psychology,* 1963, *2:*72–92.

45 Carl R. Rogers, *On Becoming a Person* (Boston: Houghton Mifflin, 1961), quote on p. 33.

Competition about the definition and goals of psychology found one focus of expression in a famous debate between Carl Rogers and B. F. Skinner conducted in 1955. In a symposium held at the annual meeting of the American Psychological Association, they discussed their contrasting views on the control of human behavior. Skinner emphasized the environmental contingencies and patterns of reinforcement that shape individual behavior. He believed that, although such conditioning is complex, it follows deterministic principles. For Skinner, the notion of free will was irrelevant and misguided. It provided a convenient fiction for maintaining the illusion that behavior originates within individuals rather than through their interaction with the environment. Rogers, in contrast, stressed the autonomy of the individual and the progressive unfolding of inherent human potential. For him, control was a dirty word, connoting coercive forces impinging upon the person from without.[46]

Both psychologists were dedicated to a science that had positive social benefits, but they disagreed on how psychology might best accomplish such goals. For Skinner, the preferred strategy was to apply the techniques of operant conditioning that had proved so powerful in the laboratory to education, commerce, and other social institutions. A true science of behavior would prove salutary by revealing the wellsprings of human action and thus provide the means consciously to manipulate behavior toward desired ends. Rogers, on the other hand, believed that such a behavioristic approach was dehumanizing, stripping people of their inalienable right to psychological self-determination. For him, a truly adequate science of psychology would assume the individual's capacity for self-fulfillment and work toward ways of maximizing its likelihood.

The Skinner–Rogers exchange sharpened the differences between behaviorism and the emerging camp of humanistic psychology. The two men perhaps realized that they were providing handy philosophical foils for each other and revisited the issues again in a marathon debate several years later.[47]

George Kelly: Science as the Model of Man

A more radical response to the reflexive imperative was to accept it as a fundamental postulate and build it into theoretical or therapeutic systems. Perhaps the most ambitious and sustained attempt to create a reflexive general theory of personality was made by George A. Kelly, director of graduate training in clinical psychology at Ohio State University from 1946 to 1965. Not easily categorized as either a humanist or a behaviorist, Kelly presented his views in a 1,200-page book, *The Psychology of Personal Constructs,* published in 1955. It developed a theory of human

46 Carl R. Rogers and B. F. Skinner, "Some Issues Concerning the Control of Human Behavior: A Symposium," *Science,* 1956, *124:*1057–1065.
47 B. F. Skinner, "Comment on Rogers," *American Psychologist,* 1974, *29:*640.

nature that talked about people as if they behaved like scientists, making hypotheses about the world and then testing them against experience. In Kelly's view, humans seek cognitive control of their world by construing events and fitting them into a framework of "personal constructs" that makes sense of them. The system carried explicit therapeutic implications: desired personality changes could be effected by adopting new personal constructs, with or without the aid of a psychotherapist.[48]

Kelly's system can be seen as his answer to the conceptual challenges he summarized in his 1939 paper "The Person as a Laboratory Subject, as a Statistical Case, and as a Clinical Client." Noting that psychologists had one theory about their own behavior as scientists and another theory about the behavior of ordinary people (i.e., their "subjects," "cases," and "clients"), he attempted to create a unified framework that would encompass both. His radical solution was to characterize people as if they behaved like scientists. That is, as if they made and tested hypotheses about the world in fundamentally the same way that scientists do in a more formal fashion. As one of his disciples put it:

> in inventing personal construct theory, [Kelly] set out to depict all persons as scientists or, for that matter, all scientists as persons. He strove to build a reflexive theory; a theory that would account for its own creation and its creator and use one language only to describe all human endeavor and confusion. By arguing that our desire to understand and anticipate is at the center of our human nature, Kelly judged science to be only a Sunday-best version of an everyday activity.[49]

Kelly's point of view derived from his varied experience as a researcher and clinician at Fort Hays Kansas State College before the war and his later efforts to build a graduate training program at Ohio State University. Although he did not align himself with any particular school of psychology, his approach was indebted to a behavioristic model of the social shaping of cognition, and he borrowed ideas eclectically from Freudian psychoanalysis, the psychodrama of Moreno, and the general semantics of Korzybski, to name a few.[50]

By removing the boundary between science and other forms of human behavior, Kelly attempted to view "man as the paradigm of the scientist – and vice

---

48 George A. Kelly, *The Psychology of Personal Constructs,* 2 vols. (New York: Norton, 1955).

49 Don Bannister, "Foreword," in Robert A. Neimeyer, *The Development of Personal Construct Psychology* (Lincoln: University of Nebraska Press, 1985), pp. xi–xii, quote on p. xi. This book analyzes the theory group that emerged from Kelly's work.

50 Robert A. Neimeyer, "George A. Kelly: In Memoriam," *History of Psychology,* 1990, *22*:3–14. For a critical perspective on Kelly's originality, see Stephen A. Appelbaum, "The Accidental Eminence of George Kelly," *Psychiatry and Social Science Review,* 1969, *3*:20–25.

versa."[51] This conceptual move stripped science of its special *separate* status in the repertoire of human action and recast it as the *general foundation* of all behavior.

The APA Self-Study of Psychology

As the expansion of psychology accelerated immediately after the war, it appeared as if the field was galloping off in all directions at once. Pride in the growth of psychology was tempered by anxiety about its implications and caused some leaders to suggest that a stocktaking was in order. In the early 1950s the Policy and Planning Board of the American Psychological Association decided that a comprehensive review of the scientific and professional status of psychology was called for and undertook a major self-study.

Formally initiated in 1952, the APA Study of the Status and Development of Psychology was divided into two components. Project A was concerned with the "methodological, theoretical, and empirical status of psychological science," while Project B explored "occupational, educational, and institutional problems." The study addressed the twin concerns over the production of scientific knowledge and the reproduction of scientific psychologists. Although research and training were inextricably bound together, the division of the study into two parts reflected the conventional distinction between scientific and professional issues.[52]

The study received a major grant from the National Science Foundation, which had been established only two years before. Under its original charter the NSF was to fund policy studies as well as research projects in an effort to contribute to national science policy. The APA, mainly through the efforts of executive secretary Dael Wolfle, had become integrated into the federal science administration and advisory apparatus based in Washington.

Project A on the scientific state of the field was directed by Sigmund Koch (1917–1996), a professor of psychology at Duke University. Conversant with the philosophy of science, which he had studied under Herbert Feigl at the University of Iowa in the late 1930s, he had a strong interest in psychological theory. During the war he had been involved as a civilian with research for the army and the Department of Agriculture, as well as teaching at Duke after receiving his Ph.D. in 1942.[53]

---

51 George A. Kelly, "Ontological Acceleration," in Brendan Maher, ed., *Clinical Psychology and Personality: The Selected Papers of George Kelly* (New York: Wiley, 1969), pp. 7–45, quote on p. 16.

52 Sigmund Koch, "Introduction," in Koch, ed., *Psychology: A Study of a Science,* vol. 1 (New York: McGraw-Hill, 1959), pp. 1–40, on pp. 5ff.

53 Lawrence D. Smith, *Behaviorism and Logical Positivism: A Reassessment of the Alliance* (Stanford, Calif.: Stanford University Press, 1986), provides a thorough account of the relations between behaviorism and logical positivism, including Koch's role in establishing the conventional account of the relationship.

In Koch's enthusiastic hands Project A grew into a massive undertaking. The original hope had been that a comprehensive review of conceptual and empirical work in contemporary psychology would yield a solid core of knowledge upon which to make further scientific advances. In the effort "to explore, not prejudge, the structure of the science," Koch cast his net widely and split the project into two divisions. Study I surveyed the conceptual and systematic foundations of psychology. It included a total of thirty-four theoretical formulations, most written by senior psychologists. Study II treated the empirical bases of the field and their place within the "matrix of scientific activity," especially in relation to the biological sciences on the one hand and the social sciences on the other.[54] The forty-two contributors to Study II were generally younger than those for the first study.

Soliciting, reviewing, and editing nearly eighty manuscripts took considerable time. In 1959 the results of Study I appeared in three volumes, followed by another set of three volumes for Study II, published in 1962–1963. A seventh volume, to be written by Koch, was to serve as a postscript to the study, but it was never published.[55] Although the volumes were widely cited in the literature and served teachers and students as useful reference works, apparently the entire series was never reviewed as a whole. Perhaps the sheer magnitude of the publication and the scope of its contents dissuaded anyone from attempting to tackle the task; maybe some were waiting for the seventh volume, advertised as the conclusion to the series. And the long publication cycle could have been a contributing factor. But even more fundamentally, the concerns embodied in the project when it was started in the early 1950s were simply not the same a decade later, when the results of the study were published. Instead of serving as a launching pad for a great leap forward in psychological research, the series became a monument to past hopes for scientific unity. To be sure, some individual chapters retained validity and importance within the boundaries of their chosen subject and served as useful benchmarks for the state of the art at the time. But considered as a whole, the entire enterprise was like the Edsel automobile introduced by Ford several years earlier – obsolete almost as soon as it appeared.

For Koch, his heroic effort to mine scientific gold from this massive excavation of knowledge ultimately proved futile. By the time the first three volumes of *Psychology: A Study of Science* appeared in 1959 his disillusionment was clear. The acid of critical analysis had corroded more than the conceptual and empirical foundations of psychology; it had also undermined his own faith in logical positivist epistemology. Koch reacted by sketching several lines of criticism that he would develop over the years.

54 Sigmund Koch, "Introduction to Study II," in Koch, ed., *Psychology: A Study of a Science,* vol. 6 (New York: McGraw-Hill, 1963), pp. 1–29, on pp. 3, 11.
55 Sigmund Koch, ed., *Psychology: A Study of a Science,* 6 vols. (New York: McGraw-Hill, 1959–1963).

A major thrust of Koch's critique was captured in his notion of "ameaningful thinking." He identified such thinking as a pernicious trend, found in modern life generally but especially prevalent in psychology. Eschewing a rigorous definition of the term, he attempted to describe it metaphorically as a syndrome in what he variously (and with a touch of humor) called "cognitive pathologistics" or "epistemopathology." In his words:

> when we think ameaningfully, there is a tendency to defend ourselves against the object of thought or inquiry rather than to understand it; to engage in a transaction with it rather than in a love affair; to use it rather than to savor it. More particularly, ameaningful thought or inquiry regards knowledge as the result of "processing" rather than of discovery; it presumes that knowledge is an almost automatic result of a gimmickry, an assembly line, a "methodology"; it assumes that inquiring behavior is so rigidly and fully regulated by *rule* that in its conception of inquiry it sometimes allows the rules totally to displace their human users.[56]

Such dependence on rules, Koch argued, is expressed in the pervasive fetish for method in psychology, in which the means of scientific research become ends in themselves. The method-fetishism of psychologists has led to a disregard for the ostensible objects of scientific investigation and has contributed to "a-ontologism," which "assumes that truth is manufactured, and manufactured more by method than by man."[57] As the distinctions between the knower and the known, the subject and the object, are blurred, psychology becomes autistic and ultimately irrelevant or trivial.

The sources of such problems were deeply rooted in the history of the discipline. Unlike other sciences, Koch stressed, *"psychology was unique in the extent to which its institutionalization preceded its content and its methods preceded its problems."*[58] Scientism had thus been incorporated into the very foundations of psychology: "From the earliest days of the experimental pioneers, man's stipulation that psychology be adequate to science outweighed his commitment that it be adequate to man."[59]

Koch realized that much of what he said sounded similar to complaints voiced by "the existentialists, the Zen Buddhists, the professional social critics, the Neo-Kantians, perhaps the Beatniks, and many other groups and individuals."[60] But he

56 Sigmund Koch, "The Allures of Ameaning in Modern Psychology," in Richard E. Farson, ed., *Science and Human Affairs* (Palo Alto, Calif.: Science and Behavior Books, 1965), pp. 55–82, quote on p. 56 (emphasis in original).

57 Ibid., p. 69.

58 Sigmund Koch, "Epilogue," in Koch, ed., *Psychology: A Study of a Science,* vol. 3 (New York: McGraw-Hill, 1959), pp. 729–788, quote on p. 783 (emphasis in original).

59 Ibid., p. 784.   60 Koch, "Allures of Ameaning," p. 81.

thought that he differed on one essential point: he viewed ameaning "as an *empirical psychological problem,* as one rooted in certain lawful regularities of human cognitive function."[61] Thus the challenge to psychology was to diagnose and treat this pathology of modern life, and thereby prove itself worthy as a truly human science. Only by healing itself could the discipline hope to heal society.

Like many of his colleagues, Koch was seeking a way to make psychology more relevant and to fulfill its humanistic promise. His approach was to pursue reforms from within, not to create a new movement or reject the enterprise itself, as others chose to do. Koch's critical work represented an effort to steer psychology back toward the central issues of human meaning explored by the humanities, to make it a science adequate to the human condition.

In contrast to the sprawling survey of psychology's expanding intellectual territory marshaled by Koch, the ground covered by the parallel study of the professional status of the field was more circumscribed. Project B of the APA self-study, on the occupational, educational, and institutional relations of psychology, was conceived as an investigation of the "environmental factors" that affected research and training in the discipline. Kenneth E. Clark (1914–), associate professor of psychology at the University of Minnesota, directed the study. Although the project gathered a wide variety of data, it focused on the factors that influenced the research productivity of psychologists.

The project revealed that the American Psychological Association and, by extrapolation, the American psychology community generally, was growing at a much faster rate than other scientific groups in the United States. After citing various measures of both absolute and relative growth in the number of American psychologists, the study explored patterns of specialization and trends in employment. They found a "startling shift" in the interests of APA members away from experimental psychology and toward clinical and other forms of applied work. Such changes in disciplinary interests were reflected in employment trends. Although colleges and universities still accounted for a majority of psychologists, academic employment was losing ground to jobs in government, business, and private practice.

The study accounted for growth in terms of societal demands: psychology was expanding in order to meet the needs expressed by various constituencies. Government and industry wanted more mission-oriented research; undergraduate and graduate students wanted more coursework and degree programs; individuals wanted more counseling and psychotherapy.

Unlike the scientific survey, this phase of the APA self-study was completed fairly quickly. In 1957 its report was released as *America's Psychologists: A Survey of a Growing Profession.*[62] It proved easier to describe the demographic char-

---

61 Ibid., p. 82 (emphasis in original).
62 Kenneth E. Clark, *America's Psychologists: A Survey of a Growing Profession* (Washington, D.C.: American Psychological Association, 1957).

acteristics of the field than to evaluate its scientific dimensions, and the volume provided a useful benchmark of changes in the interests, employment, and training of psychologists. Coincidentally, it was published shortly before the escalation of the U.S.A./U.S.S.R. space race touched off by the launching of the Sputnik satellite in 1957, which helped to spur another major increase in the federal funding of scientific research and training.

The Humanistic Movement and the Lure of Experiential Psychology

The humanistic movement in psychology represented a shift away from some of the dominant assumptions, methods, and goals of mainstream psychology. Dissatisfaction with the dogmas of Freudian theory and the narrow methods of behaviorism provided a rallying point for the emergence of the "third force" of humanistic psychology in the postwar era. Reacting against what they viewed as the mechanistic and reductionistic thrust of modern psychology, humanistic psychologists promoted a more inclusive and holistic view of human nature. In particular, they sought ways to incorporate experiential knowledge into psychological science. This, not surprisingly, was a major challenge, given the rejection of "subjectivity" in the methods and findings of mainstream psychology.

Humanistic psychologists were an eclectic lot, drawing on a variety of sources for inspiration, including existentialism and phenomenology. Having a fundamental interest in human experience from the perspective of the experiencing person, this approach privileged the uniqueness of the individual and sought ways to enhance psychological health. Hence the emphasis on "personal growth," "self-actualization," and the "realization of one's potential."[63] Explicitly concerned with values in the context of science, humanistic psychology reflected the incorporation of the idiographic approach into mainstream psychology. The institutionalization of the movement can be traced through the establishment of the *Journal of Humanistic Psychology* in 1961 and the founding of the American Association of Humanistic Psychology in 1962, and the subsequent organization of the APA Division of Humanistic Psychology in 1971.

Carl Rogers found himself among the leading figures in humanistic psychology. Somewhat to his surprise, his work was identified with the existential turn the movement took. Although he considered much research narrow, sterile, and trivial, he was unwilling to abandon his abiding faith in scientific theory and method. What Rogers did was to refurbish, not discard, the existing conceptual framework

---

63 On humanistic psychology, see Hilgard, *Psychology in America,* pp. 504–505, 786–790. See also Roy Jose De Carvalho, "A History of Humanistic Psychology (Allport, Maslow, Rogers, May, Bugental)" (Ph.D. diss., University of Wisconsin, 1988).

of psychology by privileging "personhood." Knowledge about the person rather than general laws or principles of behavior was the goal.

Abraham Maslow was another major figure in humanistic psychology. Like Rogers, he was unwilling to abandon the methods of science in psychology but sought to expand the definition of psychological science itself. He had moved away from his behavioral studies of dominance and submission in primates, conducted before the war, and turned to research on "self-actualizing" people who exhibited exceptional psychological health. His work with animals had convinced him of the importance of biological drives, but in working with humans he found it impossible to ascribe all motivations to such sources, particularly in the study of "peak experiences" wherein individuals feel a powerful sense of transcendence and enlightenment.[64]

Much of Maslow's work in the 1950s was in a philosophical vein, trying to create a normative psychology that would both account for and encourage the full development of human potential. In his 1954 book *Motivation and Personality,* he argued that "the study of crippled, stunted, immature, and unhealthy specimens can yield only a cripple psychology and a cripple philosophy. The study of self-actualizing people must be the basis for a more universal science of psychology."[65] In 1966 he produced a major statement of his views in *The Psychology of Science,* which presented a philosophical critique of objectivism and positivism in psychology.[66] In it he argued that psychology's epistemological foundations must be recast to include values as well as facts, experience as well as experiments. Understanding and empathy rather than prediction and control were to be the goals.

In a scathing review of the book, social psychologist Brewster Smith took Maslow to task for widening the definition of science to the extent that it lost all distinctive meaning. In particular, he resisted Maslow's attempt to assimilate scientific creativity to artistic creativity and his reliance on phenomenological experience as an adequate approach to explaining reality, arguing that, "as a social enterprise, science has its own agenda that is different from the arts."[67] Smith considered some of Maslow's statements about ultimate meaning and truth as a move beyond both science and art into the realm of metaphysics and theology. In general, he was sympathetic with Maslow's concern about humanizing scientific psychology, but he disagreed with the way Maslow proposed to accomplish it.

64 See Abraham H. Maslow, *The Farther Reaches of Human Nature* (New York: Viking, 1971).
65 Abraham H. Maslow, *Motivation and Personality* (New York: Harper, 1954), p. 234.
66 Abraham H. Maslow, *The Psychology of Science: A Reconnaissance* (New York: Harper & Row, 1966).
67 M. Brewster Smith, "An Ambiguous Case for Humanistic Psychology," *Science,* 1966, *153:*284–285, quote on p. 285.

Skinner and the Behavior of a Behaviorist

Although he had little regard for the humanistic movement in psychology, B. F. Skinner considered himself a humanist in the largest sense of the word.[68] And despite his aversion to subjectivity, he could not resist the temptation to turn the tools of his trade upon himself in the 1950s. During the previous decade Skinner had extended his experimental analysis of behavior beyond the context of the laboratory into a variety of real-life settings. In Project Pigeon, the Air Crib, and *Walden Two,* he attempted to demonstrate the universal validity of operant conditioning and the widespread applicability of reinforcement techniques. His system of psychology, like all such general systems, rested on the assumption that it would encompass the behavior of any organism, including humans. Thus the entire range of human behavior was amenable to behavioral analysis, in principle at least. So it was hardly surprising when Skinner began to discuss his own life history as a source of behavioral insight. The private inner dialogue he had carried on for many years, documented in his voluminous notes to himself since college days, became a reservoir of personal events that had wider significance as exemplars of his behavioral approach.

Skinner's first foray into this autobiographical mode occurred in the mid-1950s, when Sigmund Koch asked him to provide an account of his research methods for the APA study of psychological science. Using his own conduct as a researcher, he published "A Case History in Scientific Method." In a witty parody of the stuffy formalisms about scientific method common among philosophers of science as well as the general public, Skinner portrayed himself as an opportunistic individualist in the laboratory.[69]

In presenting an empirical account of a single case (his own), Skinner was criticizing not only rationalized ideas about scientific method but also the increasing domination of statistics in research design and data analysis. He argued that "statistical techniques serve a useful function, but they have acquired a purely honorific status which may be troublesome. Their presence or absence has become a shibboleth to be used in distinguishing between good and bad work."[70] He then went on to describe his own behavior as an experimenter, and derived four unformalized principles of scientific methodology from his experience:

1.   When you run onto something interesting, drop everything else and study it.
2.   Some ways of doing research are easier than others.

68 In an interview a few months before his death, Skinner stated, "I am a humanist in the sense that nothing human is alien to me." Daniel W. Bjork, *B. F. Skinner: A Life* (New York: Basic Books, 1993), p. vii.
69 B. F. Skinner, "A Case History in Scientific Method," *American Psychologist,* 1956, *11:* 221–233. Reprinted in Skinner, *Cumulative Record: A Selection of Papers,* 3d ed. (New York: Appleton-Century-Crofts, 1972), pp. 101–124.
70 Skinner, "Case History," p. 120.

3. Some people are lucky.
4. Apparatuses sometimes break down.[71]

Although Skinner's generalizations were tongue-in-cheek, they did express his sincere conviction that excessive theorizing, whether about methods, data, or results, was detrimental to scientific insight.[72]

Skinner concluded his self-analysis with a strong reflexive twist: "The organism whose behavior is most extensively modified and most completely controlled in research of the sort I have described is the experimenter himself."[73] What began as a scientific self-description in literary form in the 1950s would later blossom into a full-scale autobiographical experiment in the 1970s, when Skinner sought to describe his life history in completely behavioristic terms.

For more than a decade Skinner was occupied with his autobiography, which grew into his largest and most sustained literary production. In a series of three volumes published between 1976 and 1983, Skinner fashioned the episodes of his own life into a loosely chronological narrative.[74] Eschewing analysis of his motives or inner life, he attempted to describe his life "objectively," from the perspective of an outside observer. He used the terminology of behavioral analysis – contingencies, reinforcement, shaping – to convey the sense that his life was the result of environmental forces acting upon his biological endowment. His message was that a person's "life" was no more than a convenient shorthand label for the nexus of behavioral events impinging upon an individual. In Skinner's hands, the particular idiosyncrasies of his life, writ large, became exemplars of the universal principles revealed by behavioral psychology. Thus the tale of the scientist's behavior was transformed into a vehicle for telling the story of behavioral science. Although Skinner's inventive genius has been recognized, it has perhaps not been fully appreciated how much he applied it to the materials of his own life.[75]

Both Skinner and Rogers contributed essays to the fifth volume of *A History of Psychology in Autobiography,* published in 1967. When the two philosophical an-

---

71 Ibid., pp. 104–110.

72 Skinner's criticisms can be seen in the broader context of a general movement away from rationalistic accounts of science on the part of historians, philosophers, sociologists, and working scientists that was emerging at this time and was associated with such figures as Thomas Kuhn, Paul Feyerabend, and Michael Polanyi.

73 Skinner, "Case Study," p. 122.

74 B. F. Skinner, *Particulars of My Life* (New York: Knopf, 1976); idem, *The Shaping of a Behaviorist* (New York: Knopf, 1979); idem, *A Matter of Consequences* (New York: Knopf, 1983).

75 See Bjork, *B. F. Skinner.* In my review of the book (*Technology and Culture,* 1995, *36,* 68–69), I point out how Skinner "shared Montaigne's humanistic credo, expressed four centuries earlier, that 'each man bears the entire form of the human condition.'"

tagonists described themselves and their motivations they both emphasized inner-directedness and self-reliance, as well as an abiding curiosity about themselves. As Skinner put it: "Whether from narcissism or scientific curiosity, I have been as much interested in myself as in rats and pigeons. I have applied the same formulations, I have looked for the same kinds of causal relations, and I have manipulated behavior in the same way and sometimes with comparable success."[76]

Rogers described his own life in a similarly analytical way. He believed that his contributions to psychology occurred as he worked out the tensions between his two selves, "the sensitively subjective therapist and the hard-headed scientist." He reported realizing that "what is most personal is most general" and found that self-revelation was both personally satisfying and scientifically rewarding.[77] Although Skinner and Rogers represented the seemingly irreconcilable differences between behaviorism and humanistic psychology, in their autobiographical presentations they demonstrated a shared belief that their lives not only exemplified but warranted their claims about psychological knowledge. It was this type of reflexive response that provided some measure of epistemological common ground to postwar American psychology.

The Reflexive Turn

As psychology boomed during the 1950s, psychologists began to be aware that they had embarked on a postwar trajectory that might lead to unanticipated consequences. The large-scale self-study conducted by the American Psychological Association concluded that: "Many decisions have been made that, intentionally or not, commit the future of psychology to new directions."[78] By reflecting on themselves and their science, psychologists attempted to come to grips with a future that they had envisioned but not entirely foreseen.

The bifurcation of the content and context of scientific work reflected the dual allegiances of psychologists as they cultivated a self-image as professional scientists and as scientific professionals. Internally, the community defined itself in terms of the values of scientific research; externally, it projected an image of competent service providers. The relation between the private world of scientific research and the public sphere of professional practice became a central preoccupation. How could psychologists reconcile the tension between the values and issues

---

76 B. F. Skinner, in *A History of Psychology in Autobiography,* vol. 5 (New York: Appleton-Century-Crofts, 1967), pp. 385–413, quote on p. 407. Soon after writing the sketch, Skinner recalled: "I began to look more closely at my life as a whole. Was I not duty-bound to give a behavioristic account of myself as a behaving organism?" Skinner, *A Matter of Consequences,* p. 292.

77 Rogers, *History of Psychology in Autobiography,* quote on pp. 378, 381.

78 Clark, *America's Psychologists,* p. 25.

connected with proper scientific behavior and proper professional behavior? For some psychologists the competing demands of scientific rigor and social relevance led to the conclusion that psychology was not Janus-faced but schizoid. The challenge thus became how to overcome or heal this breach. For most of the community, however, demands for scientific rigor and social relevance were complementary rather than competing, and by embracing the reflexive nature of their discipline, psychologists were able to expand their roles as technoscientific professionals.

Psychologists completed the hermeneutic circle through various forms of self-reflection and reflexive practices by admitting subjective experience into the realm of science. Autobiographical efforts to reflect on the discipline through the lens of personal experience can be seen as a relatively simple or primitive form of reflexivity. Such forms of self-exemplification moved into more analytically sophisticated attempts to develop a psychology of the psychologist and a psychology of psychological research. Such personal and disciplinary self-reflection led some psychologists to abandon interobserver reliability as the prime criterion for valid knowledge and to embrace an epistemologically radical scientific worldview that privileged subjective knowing.

It is easier to describe than to explain the turn toward reflexivity, however. As we have seen, the varieties of reflexive practices in postwar psychology are impressive in their scope and diversity. But what accounts for them? Surely it was easier to consider such issues when the discipline was healthy and growing. With the future seemingly secure, psychologists could turn their attention inward and reflect anew on what it meant to be a science of human nature. Perhaps the very popularity of psychology encouraged psychologists in their efforts to objectify themselves and their work. Trying to convince others of the worth of their ideas and techniques may have bolstered their own belief in the technoscientific promise of psychology. There was also an element of hubris in the reflexive turn: psychology was more than scientific discipline, it was also a worldview. Americans, already the greatest consumers in history, found it easy to be consumed with themselves. Psychologists, as major players in the cultural commerce of the self, could not extricate themselves from the web they had woven about the meaning of modern life. As a discipline, psychology had held the mirror of science up to itself; perhaps it was inevitable that some of its practitioners would step through the looking glass.

Reflexivity thus provided the intellectual glue that mediated between "scientific" and "professional" psychology and held their disparate parts together. It enabled psychologists to take the kinds of everyday concerns that people seeking professional help brought to them and place them into an appropriately scientific framework. The rhetoric of the self, of experience, of subjectivity provided an idiom for translating individual problems into the universal language of science. By embracing reflexivity, wittingly or not, psychologists were able to find ways to avoid the horns of the science-versus-practice dilemma.

11

Beyond the Laboratory
Giving Psychology Away

Although the voices of professional psychologists became more numerous and audible in postwar America, they hardly constituted a monopoly on discussions about psychology and modern life. Other experts on human behavior, such as sociologists, psychiatrists, and anthropologists, offered their views to an eager public, as did various critics and commentators. The extent and variety of discourse on "the psychological society" seemed to indicate that psychology had moved to the forefront of public consciousness.[1]

The notion that Americans lived in a "psychological society" took hold rapidly in the 1950s and had become a commonplace by the 1960s. Contemporary awareness of the changing social environment was stimulated by works that plumbed the meaning of modern life. Many commentators, both inside and outside academe, framed their analysis in terms of the relations between personality and culture. Among the first studies to appear after the war was David Riesman's evocatively titled book *The Lonely Crowd* (1950), which explored the personality characteristics of people living in a mass society, particularly the shift in the location of people's moral compass from within themselves ("inner-directedness") to outside agents ("other-directedness").[2] His concern over the decline of individualism was shared by William II. Whyte, Jr., a writer for *Fortune* magazine. In *The Organization Man* (1956), he traced the rise of a set of values that privileged collective belonging, whether to a corporation, research enterprise, or other group. He argued that even scientists were embracing bureaucracy and had relinquished control over the direction of their work to their patrons in government and industry.[3]

1 See Ellen Herman, *The Romance of American Psychology: Political Culture in the Age of Experts* (Berkeley/Los Angeles: University of California Press, 1995).
2 David Riesman, with Nathan Glazer and Reuel Denney, *The Lonely Crowd,* abridged edition with new preface (New Haven, Conn.: Yale University Press, 1961; orig. pub. 1950).
3 William H. Whyte, Jr., *The Organization Man* (New York: Simon and Schuster, 1956).

From a psychoanalytic perspective, Philip Rieff, first in *Freud: The Mind of a Moralist* (1959) and then in *The Triumph of the Therapeutic* (1966), argued that the pursuit of individual self-gratification, initially legitimated by Freudian doctrine, had become an end in itself rather than a by-product of following some morally sanctioned code of conduct.[4] Former psychology graduate student Betty Friedan, after a decade of combined housewifery and freelance magazine writing, published *The Feminine Mystique* (1963), a searching critique of the cult of motherhood and the subordination of women in America.[5] Such analyses of the national psyche tended to focus on the phenomenological manifestations of social and cultural change and were less concerned with exploring the ideological and institutional structures that sustained them.

There were some attempts to lay the blame at the door of professional psychologists. Joseph Wood Krutch, author of *The Measure of Man* (1954), criticized the mechanistic assumptions prevalent in psychological theory and argued that "we have been deluded by the fact that the methods for the study of man have been for the most part those originally devised for the study of machines or the study of rats, and are capable, therefore, of detecting and measuring only those characteristics which the three do have in common."[6] Krutch's criticisms echoed earlier attacks against the intellectual arrogance of scientists, including psychologists, made by Anthony Standen in *Science Is a Sacred Cow* (1950).[7]

When viewed in historical perspective, some general themes concerning such developments have begun to emerge.[8] Only a few historians, however, have dealt with the institutional basis of this pervasive turn toward psychology in modern life. Both John Burnham and Gerald Grob have explored the social and cultural dimensions of the postwar rise of the psychological helping professions. On the expansion of psychological services, Grob argues that "[b]ureaucratic and organizational imperatives promoted the concept of adjustment, while cultural norms increasingly incorporated a preoccupation with the self."[9] Although the people performing these services differed in their doctrine, training, and occupational skills, the public easily lumped them together as professional psychotherapists. Thus, as Burnham notes, the "generic 'psychologist,'" whether in psychiatry, clinical psychology, social work, or counseling, achieved "an almost omniprofessional eminence after the mid-

4 Philip Rieff, *Freud: The Mind of a Moralist* (New York: Viking, 1959); idem, *The Triumph of the Therapeutic: Uses of Faith After Freud* (Chicago: University of Chicago Press, 1966).
5 Betty Friedan, *The Feminine Mystique* (New York: Laurel, 1983; orig. pub. 1963).
6 Joseph Wood Krutch, *The Measure of Man* (New York: Bobbs-Merrill, 1954), p. 32.
7 Anthony Standen, *Science Is a Sacred Cow* (New York: Dutton, 1950).
8 For an intellectual history, see Wilfred M. McClay, *The Masterless: Self and Society in Modern America* (Chapel Hill: University of North Carolina Press, 1994).
9 Gerald N. Grob, *From Asylum to Community: Mental Health Policy in Modern America* (Princeton, N.J.: Princeton University Press, 1991), quote on p. 108.

twentieth century" in the context of the rise of the organizational society and the developing culture of narcissism.[10]

The institutional mechanisms enabling the profound changes identified by Grob and Burnham include the experience of World War II, the rise of the federal mental health establishment, the growth of the American university system, the increasing influence of organized professional bodies, and the spread of mass media of communication. As the professional domain of psychology expanded, it spilled over into the popular realm and gave rise to an almost indescribable cornucopia of theories, techniques, and therapies offered for public consumption.

If the reflexive turn in psychology was in part a response to the epistemological issues raised by professional practice, it also proved to be an immensely useful way to connect to the popular imagination. The public, after all, had never abandoned its belief that psychology was about people and their personalities and problems. As professional psychologists began to discover the virtues of reflexivity as a scientific strategy with popular appeal, the line between the esoteric results of the laboratory and the exoteric simplifications of pop psychology became increasingly blurred.

Some psychologists began to address their work directly to the public and in so doing not only represented contemporary culture but also shaped it according to particular values. The increasing popularization of psychological knowledge in turn served to reinforce the trend toward social activism in organized psychology. The American Psychological Association, for instance, became a leading indicator of the relations between the internal self-image of the profession and public perceptions of the field.

The Formation of the Psychonomic Society

By the early 1960s the postwar growth of professional psychology had resulted in an occupational shift of historic proportions. For the first time the majority of American psychologists were employed outside traditional academic settings.[11] The trend had been unmistakable for several years. As early as 1958 Stuart Cook predicted that the term

> "psychologist" will come to mean only the person in professional practice. Given current trends, psychological scientists soon will be far outnumbered. To satisfy society's demands for needed professional services, an army of psychologists is being created; this appears to be happening faster than we can

10 John C. Burnham, "Psychology and Counseling: Convergence into a Profession," in N. O. Hatch, ed., *The Professions in American History* (Notre Dame, Ind.: University of Notre Dame Press, 1988), pp. 181–198, quote on p. 195.

11 Robert C. Tryon, "Psychology in Flux: The Academic–Professional Bipolarity," *American Psychologist*, 1963, *18:*134–143.

solve the problem of developing an integration of professional and scientific roles that is congruent with the realities of the professional job environment.[12]

His comments defined the central dilemma facing psychology: how to maintain an identity as a scientific discipline while at the same time provide a widening array of consulting services.

As clinical psychology continued to burgeon, some experimentalists began to feel like a beleaguered minority and took steps to reassert the primacy of the laboratory. By the late 1950s the continued expansion of professional psychology began to be viewed with increasing alarm by traditionalists worried that the field was straying too far from its academic roots as a laboratory discipline. The APA had grown substantially, making its ever-larger conventions even less suitable venues for serious scientific discourse. With a total membership of 18,000 at the end of the decade, the clinical division was three times the size of the experimental division.[13] In general, the concerns and activities of the association were heavily weighted toward serving the interests of practitioners.

A response to this state of affairs occurred in the summer of 1959 when an informal group of experimentalists at the University of Wisconsin discussed the idea of creating an alternative forum for scientific psychology, a subject first broached a few months earlier during a meeting of the American Association for the Advancement of Science. Desiring an organization that would focus attention on the scientific dimensions of psychology, including its teaching, they decided to approach some of their colleagues as potential members of an organizing committee, and Clifford Morgan and William Verplanck, respectively, emerged as acting chairman and acting secretary of the group. In late December 1959, at the annual meeting of the American Association for the Advancement of Science, Morgan chaired the meeting of the organizing committee, which decided to go ahead with the formation of a new organization. The eleven-member organizing committee was turned into a governing board, and Morgan and Verplanck installed as officers.[14]

In order to distance the new organization from existing groups, the name Psy-

---

12 Stuart W. Cook, "The Psychologist of the Future: Scientist, Professional, or Both," *American Psychologist,* 1958, *13:*635–644, quote on p. 643.

13 *American Psychological Association 1960 Directory* (Washington, D.C.: APA, 1960) listed 789 members of the Division of Experimental Psychology and 2,376 members of the Division of Clinical Psychology.

14 The members of the organizing committee-cum-governing board were: W. J. Brogden, W. K. Estes, F. Geldard, C. H. Graham, L. G. Humphreys, C. T. Morgan, W. D. Neff, K. W. Spence, S. S. Stevens, B. J. Underwood, W. S. Verplanck. W. R. Garner, "Clifford Thomas Morgan: Psychonomic Society's First Chairman," *Bulletin of the Psychonomic Society,* 1976, *8:*409–415. Ten of the eleven were members of the Society of Experimental Psychologists; Ernest R. Hilgard, *Psychology in America: A Historical Survey* (San Francisco: Harcourt Brace Jovanovich, 1987), p. 764.

chonomic Society was adopted. The term "psychonomic" was a little-used adjective that related in a broad sense to psychological laws or principles. It also had welcome connotations of the "nomothetic" approach (i.e., the search for generalities or universals) favored by experimentalists over the "idiographic" approach (i.e., the attempt to describe particular cases), commonly ascribed to clinicians and others more concerned with the behavior of individuals.

In early 1960 a letter of invitation to join the new society was sent to some eight hundred psychologists with "proven competence in scientific psychological research" to become charter members. The Psychonomic Society was formed "to preserve and foster high scientific standards for psychological research and the traditional academic values associated with a scientific discipline." The letter claimed that the group's sole purpose was to provide a venue for the dissemination of scientific information through an annual meeting and publications. Care was taken to point out that it was not meant to compete with or undermine the APA's Division 3 on Experimental Psychology.[15]

The first meeting of the Psychonomic Society was held in September 1960 immediately before the regular annual meeting of the American Psychological Association. At the opening session, Morgan welcomed his "fellow psychonomists and psychonomers" to the gathering.[16] After its successful inauguration, the society continued to hold annual meetings. By 1962 it had nearly nine hundred members, of whom more than 90% were also members of the APA.[17]

In 1964 Morgan launched a research journal, *Psychonomic Science,* as an independent publishing venture. It was designed to promote communication among researchers by providing quick publication of short scientific articles. Morgan soon added three related journals: *Psychonomic Monograph Supplements* (1965), *Perception and Psychophysics* (1966), and *Behavior Research Methods and Instrumentation* (1969). In 1968 he offered the entire publishing enterprise to the Psychonomic Society, along with an endowment for its operation. The transfer occurred in 1970, and Morgan remained as publications manager for the society.[18]

15 The letter is cited and discussed in Hilgard, *Psychology in America,* pp. 764–765, 847.
16 Morgan's hope that "psychonomic" and related terms such as "psychonomer" and "psychonomist" would gain currency in describing experimental psychologists and their work was not fulfilled; Garner, "Clifford Thomas Morgan," p. 414.
17 By 1981 the Psychonomic Society had 3,000 members, approximately 80% of whom were also APA members; Hilgard, *Psychology in America,* p. 765.
18 Morgan was a highly successful textbook author and had become independently wealthy from the sales of his *Introduction to Psychology* (1956). His editorial skills were matched by his business acumen about the publishing industry. (Garner, "Clifford Thomas Morgan," pp. 412–415.) Since this section was written a detailed account of the origins of the society has appeared: Donald A. Dewsbury and Robert C. Bolles, "The Founding of the Psychonomic Society," *Psychonomic Bulletin & Review,* 1995, 2:216–233.

The Expansion of Clinical Psychology

As experimentalists dedicated to a nomothetic science of psychology took steps to sustain their own institutional subculture, the American Psychological Association continued to focus on the professional issues raised by the growth of clinical psychology. Shortages of trained personnel remained a problem, despite the support of the Veterans Administration and the Public Health Service for graduate training and internships.

Clinical psychologists took increasing interest in psychotherapy, and more and more of them established private practices. Licensing and certification were hotly debated as states took steps to control the practice of psychology. There was some friction with the psychiatric profession as clinical psychologists continued to encroach on traditional medical turf. Policing the borders between psychology and medicine raised many problems because many internship and employment opportunities for psychologists were afforded by hospitals and other medical facilities. The battle was joined on intellectual grounds as well, as the medical disease model for psychological disorders came under criticism.[19]

Some called for greater autonomy and freedom from medical control. Among the loudest voices was that of George Albee, a clinical psychologist at Western Reserve University and former employee of the APA central office. In 1963, as president of the Ohio Psychological Association, he issued a "Declaration of Independence for Psychology" that called for the establishment of independent psychological centers. Such facilities would deliver mental health services as well as provide training opportunities.[20]

Albee had been the author of an influential report, *Mental Health Manpower Trends* (1959), that had helped to catalyze support for the National Community Mental Health Centers Act, passed in 1963. This legislation was designed to expand the mental health care delivery system by creating local centers to serve every geographical area of the country. Although the community mental health centers were not under the exclusive control of psychologists, they did provide a new venue where clinical psychologists could exercise their professional skills.[21]

19 The history of certification and licensing is beyond the scope of this study; for an entry into the issues, see Erasmus L. Hoch, "The Profession of Clinical Psychology," in B. B. Wolman, ed., *Handbook of Clinical Psychology* (New York: McGraw-Hill, 1965), pp. 1427–1442. For a classic critique of the medical model, see Thomas S. Szasz, *The Myth of Mental Illness* (New York: Hoeber-Harper, 1961).

20 George Albee, "A Declaration of Independence for Psychology," *Ohio Psychologist,* 1964 (June), n.p. See also George Albee, "Letters," in T. S. Krawiec, ed., *The Psychologists,* vol. 3 (Brandon, Vt.: Clinical Psychology Publishing Co., 1978), pp. 3–59.

21 George W. Albee, *Mental Health Manpower Trends* (New York: Basic Books, 1959). See also Murray Levine and Adeline Levine, *A Social History of Helping Services* (New York: Appleton-Century-Crofts, 1970); Murray Levine, *The History and Politics of Community Mental Health* (New York: Oxford University Press, 1981).

In 1965 a follow-up to the Boulder conference on graduate training in clinical psychology was held. The Conference on the Professional Preparation of Clinical Psychologists involved fifty-eight participants representing a wide spectrum of the profession. Sponsored by the APA with funds from the National Institute of Mental Health, it was held in Chicago in late summer. In his opening remarks, incoming APA president Nicholas Hobbs argued that the profession should make itself useful to the emerging "Great Society" programs championed by the presidential administration of Lyndon B. Johnson rather than pursuing the practice of individual psychotherapy, which was limited to the fortunate few who could afford it. He stated, "our central concern should be to identify ways in which clinical psychologists can be trained to be of maximum social use in a national program committed to the fullest development of people."[22] The "public clamor for psychological services" was an ever-present theme; the organizers admitted that the conference was convened because "the profession seems to have stimulated a demand it cannot meet."[23]

With a slight but significant terminological change from "scientist-practitioner" to "scientist-professional," the conference endorsed the model first articulated at the Boulder conference in 1949. Research training, internship experience, and training in psychotherapy were all stressed as essential components of a good doctoral program. Alternative conceptions of the role of the clinical psychologist were also discussed, especially the professional-psychologist model. No consensus was reached on the advisability of developing a program devoted to the training of full-time practitioners. There was some agreement, however, that such an emphasis could be accommodated under a more diversified scientist-professional training program.

In the face of some reluctance by university psychology departments to expand their clinical training programs in response to increased demand, the conference encouraged individual institutions to explore and experiment on their own. In the concluding statement, psychologists once again took refuge in the virtues of laissez-faire: "psychology ought to keep an open mind, letting the results speak for themselves."[24]

Psychology Goes Public

As debates over policy and technical issues were joined within the professional community, psychology was also gaining in public visibility. In the mid-1950s the field achieved its first exposure on network television. Joyce Brothers (1929–) launched

22 Nicholas Hobbs, "Opening Address," in Erasmus L. Hoch, Alan O. Ross, and C. L. Winder, eds., *Professional Preparation of Clinical Psychologists* (Washington, D.C.: American Psychological Association, 1966), pp. 1–6, quote on p. 4.

23 Hobbs, "Background of the Conference," in ibid., pp. 7–9, quote on p. 8.

24 Hobbs, "Summary of Conference," in ibid., pp. 79–93, quote on p. 93.

her television career and by the end of the decade had her own show, dispensing advice as a consulting psychologist. Her professional trajectory was unusual, to say the least, but perhaps indicative of the growing role of popular psychology.

After earning her undergraduate degree in psychology at Cornell in 1947, Brothers entered the Columbia University graduate program. She completed her Ph.D. in 1953, with a dissertation in physiological psychology. Although she had previously served as an instructor at Hunter College, after graduation she gave up academic life in order to stay at home to rear her daughter. The burden of financing her husband's medical education led her to audition for *The $64,000 Question,* a popular television quiz show on which contestants competed for cash prizes by answering questions concerning unusual or arcane subjects. Selected for a show featuring the sport of boxing, Brothers made herself an expert by mastering a twenty-volume encyclopedia on the topic. She recalled later, "I had good motivation because we were hungry."[25]

Brothers first appeared on the quiz show in the fall of 1955 and built up her winnings through successive rounds of questions until she won the $64,000 question in December. She became the second person and the only woman to win the grand prize. Two years later she appeared on the show's successor, *The $64,000 Challenge,* which featured a panel of experts who tried to stump the contestant. She won again, successfully competing against a group of seven ex-boxers. She became one of the biggest winners in television quiz show history, with total earnings of $134,000.

Brothers made other appearances on television and drew the attention of producers at the National Broadcasting Company. She was offered a four-week contract to host an afternoon show "given over to counseling and advice on love, marriage, sex, and child-rearing."[26] It was an immediate success, and soon she had her own nationally syndicated show, "Dr. Joyce Brothers." Her dignified manner and matter-of-fact style enabled her to discuss subjects such as menopause or impotence that had previously been taboo on the air.

Brothers's career on national television continued through the 1970s, and she became a one-person media conglomerate, diversifying into various radio and television formats, both taped and live, as well as the print media, with newspaper and magazine columns and a number of popular books. Careful to distinguish her brand of "psychological uplift" from professional psychotherapy, Brothers considered her role to be a mediator between the public and the psychological literature.[27] As one observer noted, "She not only brings the lessons of psychology, for better or worse, to millions of Americans, she *is* psychology to millions of Americans."[28]

---

25 "Brothers, Joyce," *Current Biography,* 1971, pp. 66–68, quote on p. 66.

26 Ibid., p. 67.   27 Ibid.

28 "Brothers, Joyce," *Contemporary Authors,* New Revision Series, vol. 13, pp. 75–76, quote on p. 76.

The spectacular media career of Joyce Brothers is perhaps the most outstanding example of the popularization of psychology after World War II. Her success, replicated in smaller ways by countless others, contributed to the flood of psychological information available to the lay public on the airwaves and in newsstands and bookstores.[29]

At the same time Brothers was breaking into television, Timothy Leary (1920–1996) was working as director of psychological research at the Kaiser Foundation Hospital in Oakland, California. He received his undergraduate education during World War II, spending his first year at Holy Cross College in his home state of Massachusetts, the next year and a half at the U.S. Military Academy at West Point, and then finishing at the University of Alabama with a bachelor's degree in 1943. For the remainder of the war he worked as a psychologist in an army hospital in Pennsylvania. Returning to graduate school, he earned a master's degree at the University of Washington (1946) and a doctorate at the University of California (1950). Upon graduating he became an assistant professor in the Berkeley psychology department, where he produced the manual *Multilevel Measurement of Interpersonal Behavior.*[30]

After five years at Berkeley, Leary became director of psychological research at the Kaiser Foundation Hospital in nearby Oakland, from 1955 to 1958. Here he continued his work on personality theory and psychotherapy and devised a personality test that became widely used by government and private agencies. During this period he published two major works, *Interpersonal Diagnosis of Personality* and *The Prediction of Interpersonal Behavior in Group Psychotherapy.*[31] After his wife's premature death in 1958, Leary underwent a personal crisis. He took leave from his job and traveled to Spain with his two young children. In the wake of an episode of delirium induced by a short but severe illness, he had a life-clarifying insight into his own personality.

According to his own account, he was determined not to fall back into old patterns of behavior. He returned to the United States in 1959 to a position as lecturer at Harvard in the Center for Research in Personality. Borrowing from decision theory and

---

29 The psychologist-as-celebrity is among the themes that were explored by John C. Burnham, *How Superstition Won and Science Lost: Popularizing Science and Health in the United States* (New Brunswick, N.J.: Rutgers University Press, 1987), p. 238.

30 Timothy Leary, *Multilevel Measurement of Interpersonal Behavior* (Berkeley, Calif.: Psychological Consultation Service, 1956).

31 Timothy Leary, *Interpersonal Diagnosis of Personality: A Functional Theory and Methodology for Personality Evaluation* (New York: Ronald, 1957); Timothy Leary and Hubert S. Coffey, *The Prediction of Interpersonal Behavior in Group Psychotherapy* (Berkeley, Calif.: Beacon House, Psychodrama and Group Psychotherapy Monographs No. 28, 1955).

transactional analysis, Leary "began to evolve and enunciate the theory of social interplay and personal behavior as so many stylized games, since popularized by Dr. Eric Berne in his bestselling book *Games People Play* [1964], and to both preach and practice the effective but unconventional new psychiatric research technique of sending his students to study emotional problems such as alcoholism where they germinate, rather than in the textbook or the laboratory."[32]

Leary first began to experiment with psychoactive drugs in 1960, when he ingested psylocybin mushrooms in Mexico. Back at Harvard he also tried mescaline, and then lysergic acid diethylamide (LSD). Convinced that he had stumbled upon a valuable tool for the exploration of human consciousness, he continued his personal experimentation with a small circle of colleagues, students, and friends. He also conducted controlled experiments involving drugs with volunteer inmates from a local prison and with individuals suffering from alcoholism, schizophrenia, and other psychological maladies. Although the results of such studies were promising, they proved increasingly controversial, and the Harvard administration declined to renew his contract after 1963.

One of Leary's Harvard colleagues, Richard Alpert, became a fellow-traveler in exploring psychedelic consciousness. Looking back, he described a frustrating lack of connection between his work and life:

> My colleagues and I were 9 to 5 psychologists: we came to work every day and we did our psychology, just like you would do insurance or auto mechanics, and then at 5 we went home and were just as neurotic as we were before we went to work. Somehow, it seemed to me, if all of this theory were right, it should play more intimately into my own life. . . . Something was wrong. And the something wrong was that I just didn't know, though I kept feeling all along the way that somebody else must know even though I didn't. The nature of life was a mystery to me. All the stuff I was teaching was just like little molecular bits of stuff but they didn't add up to a feeling anything like wisdom. I was just getting more and more knowledgeable.[33]

Like Leary, Alpert left Harvard and academic life. After several years of experimentation with LSD and other psychoactive drugs, he eventually discovered the enlightenment he was searching for during a trip to India. Changing his name to Baba Ram Dass, Alpert became a popular American guru.

Meanwhile, from his academic seat Leary leapt nimbly onto the public stage. He continued experimenting with psychedelic drugs and shifted his concern from

32 "Leary, Timothy," *Current Biography,* 1970, pp. 244–247, quote on p. 245; original in *Playboy* interview, September 1966. On Berne's work, see Kurt Back, *Beyond Words* (New York: Pelican Books, 1973; orig. pub. 1972), pp. 147–148.

33 Baba Ram Dass, "Journey: The Transformation: Dr. Richard Alpert, Ph.D., into Baba Ram Dass," in *Be Here Now* (New York: Crown, 1971), pp. 9–48, quote on p. 12.

scientific legitimation to a general public appeal to "turn on, tune in, and drop out." In 1966 he formed the League for Spiritual Discovery, conceived as a religious body dedicated to the expansion of consciousness through mind-altering drugs. Around the same time Leary began to compare LSD to the microscope as an investigative tool, arguing that the mind-expanding drug was as essential to psychology as the microscope was to biology. Partly for legal reasons, he took refuge in the rhetoric of religion and translated his earlier scientific concern with personality development into a mystical quest for personal enlightenment on a mass scale. The transformation from respected scientist to counterculture guru was so complete that Leary's origins as an academic psychologist have become obscured. Nonetheless, whatever the revolutionary overtones of his message, it expresses the dominant postwar vision of psychology as an instrument of personal and group transformation.

Another psychologist whose work was greeted with greater popular than professional acclaim was Arthur Janov, originator of "primal therapy." In 1970 Janov published a book, *The Primal Scream,* that described his innovative therapeutic method. In essence, patients were urged to scream like "what one might hear from a person about to be murdered" in order to confront their psychological pain and thereby relieve it.[34]

Janov had received his Ph.D. in clinical psychology in 1960 from Claremont Graduate School and had been on the staff of the Psychiatric Department of the Los Angeles Children's Hospital for several years. He described his discovery of the primal scream as a case of serendipity that occurred when he pondered taped sessions with two patients on which screaming seemed to have a healing effect.

Reviewers of the volume were generally unimpressed. One noted that the technique had much in common with various forms of "abreaction" in psychoanalysis, although Janov made no references to such literature.[35] Another reviewer suggested that Janov had violated the APA ethical code calling for "modesty, scientific caution, and due regard for the present limits of knowledge" in presenting his work to the public.[36] Despite such negative critical responses, the book was widely circulated and reprinted in a paperback edition, and Janov went on to a successful career as a well-known popularizer. His books, like those of other psychologists who appealed directly to the public, had the initials "Ph.D." prominently displayed after his name.

34 Arthur Janov, *The Primal Scream: Primal Therapy: The Cure for Neurosis* (New York: Putnam, 1970).

35 Anthony Storr, "A Short Course in Brainwashing" (review of Janov, *The Primal Scream*), *Book World,* 1970, *4*:10.

36 Gerald A. Ehrenreich, "Neurosis Cured, Claims Psychologist" (review of Janov, *The Primal Scream*), *Contemporary Psychology,* 1971, *16*:220–221, on p. 221.

Another flurry of popular interest in psychology was generated in the unlikely field of computer programming in the mid-1960s, when an experiment raised some interesting questions about psychotherapy and inadvertently revealed the instrumentalist bias inherent in Carl Rogers's client-centered approach. In an attempt to create a program that would enable a computer and a person to communicate in natural language, Joseph Weizenbaum, a computer scientist at Massachusetts Institute of Technology, developed ELIZA, named for the main character in George Bernard Shaw's play *Pygmalion.* The initial script imitated the responses a nondirective psychotherapist might use in a counseling session. The session was interactive, and took place on-line. Thus a typical conversation might begin (computer response in small capitals):

Men are all alike.
IN WHAT WAY?
They're always bugging us about something or other.
CAN YOU THINK OF A SPECIFIC EXAMPLE?
Well, my boyfriend made me come here.
YOUR BOYFRIEND MADE YOU COME HERE.
He says I'm depressed much of the time.
I AM SORRY TO HEAR YOU ARE DEPRESSED.[37]

Weizenbaum intended his use of Rogerian language as a parody, chosen because it was relatively easy to program, not for its intrinsic significance. His aim was to demonstrate the possibility of natural-language programming, not to automate the counseling process. His invention soon caught the attention of others outside of computer science, and a group of researchers at Stanford began to explore its potential as a psychotherapeutic tool. They stressed the similarities between the program and a live therapist, arguing that its essential features emulated the interpersonal relationship that develops between therapist and client. The authors acknowledged that there were some technical hurdles to overcome, such as improving the flexibility of the program, but were optimistic in projecting long-term goals such as developing "an internal cognitive model of the person during on-line communication" that would further customize the feedback provided by the computer. Although the method had not yet undergone any clinical trials, the Stanford group suggested that if it were successful it would make psychotherapy available on a mass scale. Mental hospitals and psychiatric

---

37 Joseph Weizenbaum, "ELIZA – A Computer Program for the Study of Natural Language Communication between Man and Machine," *Communications of the ACM,* 1966, *9:* 36–45, quote on p. 36.

clinics could overcome chronic shortages of therapists by recourse to computer systems.[38] Weizenbaum agreed that his program raised a number of interesting psychological implications, but he believed that the technical problems of computers would delay any practical applications for a long time.[39]

Even at the outset, Weizenbaum saw that his program raised a fundamental ethical issue. How should people assign credibility to the output of a machine? ELIZA showed how easy it was "to create and maintain the illusion of understanding, hence perhaps of judgment deserving of credibility."[40] Such concerns, however, were hardly addressed as the experiment took on a life of its own. What had begun as an exploration of natural-language programming was transformed into a demonstration of automated psychotherapy. ELIZA and similar programs achieved the status of parlor games in some academic circles and became just another passing fad among intellectuals.[41] Looking back, Weizenbaum was amazed by the strong emotional involvement people had with the program and how seriously his program was taken by some psychotherapists and was dismayed by the lack of attention paid to the ethical ramifications of his work.[42]

The ease with which client-centered therapy, devised by one of the heroes of humanistic psychology, was translated into a computer program demonstrated its reliance on concepts of feedback (i.e., the use of recursive and reiterative verbal responses by the therapist). Such concepts, generated in the technocratic environment of World War II, played a major role in breaking down distinctions between mechanisms and organisms.[43] It is ironic that one of the leading advocates for a person-centered psychology could find his work so readily expressed in machine language.

---

38 Kenneth Mark Colby, James B. Watt, and John P. Gilbert, "A Computer Method of Psychotherapy: Preliminary Communication," *Journal of Nervous and Mental Disease*, 1966, *142:*148–152.

39 Joseph Weizenbaum, "Contextual Understanding by Computers," *Communications of the ACM*, 1967, *10:*474–480.

40 Weizenbaum, "ELIZA," pp. 42–43.

41 In the mid-1970s Carl Sagan echoed earlier hopes for automated therapy: "I can imagine the development of a network of computer psychotherapeutic terminals, something like arrays of large telephone booths, in which for a few dollars a session, we would be able to talk with an attentive, tested, and largely nondirective psychotherapist." Sagan, "In Praise of Robots," *Natural History 84*(1) (1975): 8–12, 14–16, 20, quote on p. 10. See also A. K. Dewdney, "Artificial Insanity: When a Schizophrenic Program Meets a Computerized Analyst," *Scientific American 252*(1) (1985): 14–20, 120.

42 Joseph Weizenbaum, *Computer Power and Human Reason: From Judgment to Calculation* (San Francisco: Freeman, 1976), pp. 5–8.

43 See Steve Heims, "Encounter of Behavioral Sciences with New Machine–Organism Analogies in the 1940's," *Journal of the History of the Behavioral Sciences*, 1975, *11:* 368–373.

Another psychologist whose work became popular during the 1960s was Erik Erikson (1902–1994). He found inspiration for his work in the materials of his own life, and was able to generalize a theory of identity from them. Because of his unorthodox background and European psychoanalytic training, Erikson hardly qualified as a typical professional psychologist. However, despite the lack of any formal academic degrees, he was able to become a member of the American Psychological Association in 1950. His remarkable career demonstrated that marginality in professional terms was hardly an obstacle to cultural renown in an era eager for self-knowledge. Erikson's stress on identity formation and the life cycle struck a responsive chord. His first book, *Childhood and Society* (1950), recast psychoanalytic theory in developmental terms and drew upon anthropological as well as clinical research.[44]

During the 1960s Erikson's lectures at Harvard were packed. Students flocked to the teacher whose concept of identity formation seemed to articulate their experience, and the term "identity crisis" became an idiomatic expression. Continuing his work at the intersection of culture, personality, and history, Erikson presented case studies of Martin Luther and Mahatma Gandhi as exemplars of the processes of psychological development.[45]

Erikson, like his psychoanalytic predecessors Freud and Jung, used his own experiences as a seedbed for his psychological theory. He called his technique "disciplined subjectivity." Ignoring artificial measures of scientific "proof" like statistics, such authors asked that their theories be validated against the personal experience of their readers. In Erikson's case, his entire oeuvre was an exercise in self-exploration, as the challenges of biographical understanding became inextricably bound up in the living of his own life.[46] The immense popular appeal of the life stories told by Erikson suggested that they were emblematic of the modern condition.

Another indication of popular interest in psychology was provided by the monthly magazine *Psychology Today,* begun in 1967. Within a few years its circulation totaled over one million, indicating the existence of a mass market for "relatively authoritative popularization."[47] Publishing articles written by respected researchers as well as science writers, it deliberately sought to avoid the excesses of sensa-

---

44 Erik H. Erikson, *Childhood and Society* (New York: Norton, 1950; 2d ed. 1963).
45 Erik H. Erikson, *Young Man Luther* (New York: Norton, 1958); idem, *Gandhi's Truth* (New York: Norton, 1969); idem, *Identity: Youth and Crisis* (New York: Norton, 1968). See also Robert Coles, *Erik H. Erikson: The Growth of His Work* (Boston: Little, Brown, 1970).
46 See the forthcoming biography by Lawrence J. Friedman, *Becoming Erik Erikson: A Life* (New York: Scribners).
47 Burnham, *How Superstition Won*, p. 106.

tionalism. As one professor noted, *Psychology Today* "presented psychology as scientific, intelligent, idealistic, and relevant. It treated psychology with interest, respect, and enthusiasm, everything that the public and the discipline could want. Even for those psychologists who were uncomfortable with the notion of popularizing psychology, the magazine was hard to resist. Many of us subscribed to it and recommended it to our students."[48] But after a few years of maintaining a highbrow approach, *Psychology Today* became more sensationalized, placing an increased emphasis on nudity, sex, and similar topics. Psychology was seen as a source of news, not a sign of scientific progress, and the journal began to portray the field as "a life style, not an intellectual stance."[49] *Psychology Today* continued as a commercial venture through the 1970s, its pages reflecting the public appetite for psychological information and the ambiguous success of professional psychologists in feeding it.[50]

The 1969 Meeting of the APA

As individual psychologists found increasing legitimation for their work from American society, the professional community was undergoing a similar collective transformation. It became explicit at the 1969 annual convention of the American Psychological Association, which proved unique in the history of the organization. For the first time the program did not reflect the scientific preoccupations of the day, with presentations of research findings and discussions of theoretical implications in a host of separate specialties. Instead it was organized explicitly around a common theme, a theme that expressed the growing concern of the profession about their social role. The theme was "Psychology and the Problems of Society."

The papers dealt with psychological perspectives on the leading social problems of the day: urban blight, disadvantaged children, racism, violence, campus revolt. Along with discussion of specific problems, the program included more general consideration of the role of psychology in relation to social change.[51]

The special program was developed in response to the concerns of many psychologists over escalating social problems and the ineffectiveness of civic responses.

48 Gregory Kimble quoted in Gary R. Vandenbos, "The APA Knowledge Dissemination Program: An Overview of 100 Years," in R. B. Evans, V. S. Sexton, and T. C. Cadwallader, eds., *The American Psychological Association: A Historical Perspective* (Washington, D.C.: APA, 1992), pp. 347–390, on p. 374.

49 Burnham, *How Superstition Won,* pp. 115, 238.

50 In an effort to exert control over the public understanding of psychology, the APA, in a highly controversial move, bought *Psychology Today* in 1983 and published the magazine for five years. The association lost more than $15 million before divesting itself of the magazine. Vandenbos, "APA Knowledge Dissemination," p. 376.

51 Frances F. Korten, Stuart W. Cook, and John I. Lacey, eds., *Psychology and the Problems of Society* (Washington, D.C.: American Psychological Association, 1970).

This general concern was catalyzed by the 1968 Democratic National Convention, held in Chicago. Many people were disturbed by what they perceived as the cynical manipulation of the nominating process. This, combined with the spectacle of demonstrators being beaten by police outside the convention hall, prompted the APA to move its 1969 convention site from Chicago to Washington.

Additional calls for action by the Ad Hoc Committee of Psychologists for Social Responsibility found support from other groups, including the brand-new American Psychologists for Social Action as well as the venerable Society for the Psychological Study of Social Issues. The APA appointed a study group, which recommended that the 1969 convention program be arranged "to crystallize and make visible the concern of psychologists about current social problems, to inform A.P.A. members about such problems, and to suggest practical possibilities for future, more effective contributions by A.P.A. members to the solution of these problems."[52]

What was notable about the program was not its content but its official sponsorship. After all, psychologists had been expressing and acting on their concern with social problems for decades, individually as well as collectively.[53] Indeed, much of the growth of the field after World War II was due to the profession's embracing of increased social relevance. What was different about the 1969 meeting was that it had the imprimatur of the country's largest association of professional psychologists. This formal endorsement by the APA suggested that the profession was finally secure enough to acknowledge the activist impulse that had shaped its development for the last quarter-century.

In the spirit of the program, APA President George A. Miller provided the keynote in his address on "Psychology as a Means of Promoting Human Welfare." An experimental psychologist by trade, he had earned his Ph.D. at Harvard in 1946 as a member of the Psycho-Acoustic Laboratory. Miller took his title from the bylaws of the APA, which stated its intertwined aims "to advance psychology as a science and as a means of promoting human welfare by the encouragement of psychology in all of its branches in the broadest and most liberal manner."[54] Miller was well aware that many took the growth and success of applied psychology as evidence that psychologists were making important contributions to human welfare. He urged his listeners, however, to consider a more radical interpretation of the phrase.

Miller argued that it was the potential of *scientific* psychology to understand and

---

52 Stuart W. Cook, "Introduction," in ibid., pp. ix–xii, quote on p. xi.
53 See, for example, Benjamin Harris, Rhoda K. Unger, and Ross Stagner, eds., "50 Years of Psychology and Social Issues," *Journal of Social Issues,* 1986, *42*(1), 1–227.
54 George A. Miller, "Psychology as a Means of Promoting Human Welfare," *American Psychologist,* 1969, *24:*1063–1075; reprinted in Korten, Cook, and Lacey, *Psychology and the Problems of Society,* pp. 5–21, quote on p. 7.

control behavioral phenomena that constituted the true means for promoting human welfare. On this view, psychology could be "one of the most revolutionary intellectual enterprises ever conceived by the mind of man."[55] Miller went on to state: "The heart of the psychological revolution will be a new and scientifically based conception of man as an individual and as a social creature."[56] The notions of reinforcement and behavioral control had already escaped from the laboratory and become a part of common discourse. In an effort to lessen concern over the negative connotations of control, Miller suggested an emphasis on understanding and prediction instead.

Because the demand for psychological services seemed practically limitless, Miller encouraged people to become their own psychologists. He suggested that psychologists should act less as experts and try to help people learn to help themselves. The role of the psychologist would then be to transmit scientifically valid knowledge to those who needed it – that is, to everyone. Thus "giving psychology away" should be the prime function of the professional community.

Expressed in the popular rhetoric of revolution, Miller's speech struck a chord. What could be more altruistic and practical at the same time than to make psychological expertise widely and freely available to everyone? Miller also reasserted the primacy of the scientific quest in maintaining that the conceptual changes wrought by psychology were more significant than whatever practical techniques could be derived from them.

The year after his term as APA president Miller coedited a semischolarly survey of psychology with Kenneth Clark for the National Academy of Sciences. After reviewing developments in various research areas, the book ended with a series of recommendations concerning national support for psychology. Its concluding paragraph read:

> Psychology is a field that is easily susceptible to change. It has not suffered the rigidity of many other disciplines: it has accommodated itself to the changes in views and values of its members. In its early days, psychologists developed a substantial base of laboratory and experimental work that has increased the rigor and objectivity of the work of all psychologists. During the past thirty years it added an emphasis on the application of psychology to the problems of individuals. We believe that during the next generation psychology will change again. We are willing to predict that the new area of emphasis will relate to the problems of groups and societies. We not only accept this new direction but welcome it and call upon our fellow psychologists and our supporters to assist in accelerating it. As a science fitting into the total spectrum of the behavioral and social sciences, and as a profession dedicated to human welfare,

55 Ibid., p. 8.   56 Ibid., p. 9.

psychology in this and in the next generation merits the enthusiastic support of our society.[57]

Such statements, articulated and endorsed by one of the leading members of psychology's scientific elite, suggested that the technoscientific aims of pure and applied psychology were now indistinguishable. Over the preceding quarter-century the alliance for scientific professionalism had successfully united psychologists in a search for the self that provided a language for Americans to express their hopes and fears about themselves and others.

57 Kenneth E. Clark and George A. Miller, eds., *Psychology* (Englewood Cliffs, N.J.: Prentice-Hall, 1970), quote on p. 140.

Interlude V

As psychology boomed in America during the 1950s Edwin Boring was flourishing. He continued working his usual schedule after reaching normal retirement age at the beginning of the decade and finally achieved some measure of personal satisfaction in his professional accomplishments. He had found an important role for himself as psychology's great communicator.

Within the profession, Boring continued to add his voice to discussions about psychology's past, present, and future. He was called upon to provide counsel and advice to the American Psychological Association on important matters of policy and procedure. For instance, in 1954, at the height of the hysteria over Communist subversion whipped up by Senator Joseph McCarthy, Boring was appointed chair of the APA Committee on Freedom of Enquiry, which was charged with the task of exploring ways to protect academic freedom.

In 1955 the American Psychological Association launched a new journal devoted to book reviews. It was christened *Contemporary Psychology* and Boring served as editor. He sought a cosmopolitan style modeled after the *New Yorker* or the *Saturday Review*. Known as *CP* (in a sly play on the well-known abbreviation for the Communist Party), the new bimonthly journal provided a platform for its editor to pronounce on all things psychological in a column entitled "CP Speaks."

The journal was an immediate success. In addition to providing timely and in-depth reviews of the burgeoning literature of psychology, it fostered discussion of professional as well as technical issues. Boring's distinctive commentary, combined with his vigorous attempt to promote ad verbum analysis rather than ad hominem comments, made *CP* a lively forum. Unlike other publication outlets dedicated to specialist concerns, it endeavored to serve a general audience and enhance disciplinary cross-fertilization – no mean challenge, as the APA registered more than ten thousand members.

Boring also emerged as a visible public spokesman for psychology in the new medium of television. In the fall of 1956, his last year before retirement, he was approached by Boston's public television station, WGBH, with a proposal to televise his introductory psychology class lectures. The Ford Foundation, as part of its programmatic focus on behavioral science, paid half of his professional salary. Eager as always for the challenges of a new communications medium, Boring launched into the project with his usual intensity, and produced a total of thirty-eight half-hour programs. Boring found it fairly straightforward to adapt his polished Psychology 1 lectures to the new format. After having taught the course for more than thirty years at Harvard, he knew the material well and found it easy to talk to the camera "in a friendly, enthusiastic, paternal manner."[1] Now the literate public could obtain the same introduction to modern psychology that generations of Harvard undergraduates had experienced.

"Mr. Psychology" became professor emeritus in 1957 but continued as editor of *Contemporary Psychology* until 1961. Somewhat piqued at not being considered for another six-year stint as editor, Boring had to admit ruefully that his age was becoming a factor. He had lived long enough that the death of friends and professional associates was not an uncommon occurrence. As a natural outgrowth of his historical interests Boring had served as the psychology community's chief necrologist for decades, wrapping his subjects in a verbal shroud for their journey to the scientific Valhalla. He wrote a few lengthy obituaries himself, beginning with Titchener's in 1927, and also arranged for others to be written. He periodically published lists of deceased psychologists with their dates of birth and death.[2] After Robert Yerkes and Lewis Terman both died in 1956, Boring published memoirs on their scientific lives. He also revised his own autobiography. In a volume entitled *Psychologist at Large,* published in 1961, he produced something of a summary statement of his life.[3] The book contained an updated version of his autobiography, some reprinted essays and editorials, a few selections from his voluminous correspondence, and a bibliography of his writings.

By the early 1960s the history of psychology was emerging as a scholarly specialty. Not surprisingly, Boring's patronage was sought by those interested in cultivating the field, and his work was viewed as foundational. In 1963 a number of his papers were collected into a volume entitled *History, Psychology, and Science.*[4] It was edited by Robert I. Watson, a clinical psychologist, and Donald T. Camp-

1 Edwin G. Boring, *Psychologist at Large* (New York: Basic Books, 1961), pp. 76–77.
2 Edwin G. Boring, "Psychological Necrology (1903–1927)," *Psychological Bulletin,* 1928, *25:*302–305, 621–625; Edwin G. Boring and Suzanne Bennett, "Psychological Necrology (1928–1952)," *Psychological Bulletin,* 1954, *51:*75–81.
3 Boring, *Psychologist at Large.* This volume contains an updated version of Boring's autobiography from *A History of Psychology in Autobiography,* vol. 4 (Worcester, Mass.: Clark University Press, 1952), pp. 27–52.
4 Edwin G. Boring, *History, Psychology, and Science: Selected Papers,* ed. Robert I. Watson and Donald T. Campbell (New York: Wiley, 1963).

bell, a social psychologist, who wanted to showcase Boring's contributions to the history and psychology of science. In 1965 the Division for the History of Psychology was organized in the APA, counting more than two hundred charter members. Boring was honored by election as its first president.

In the summer of 1968 Harvard's Stillman Infirmary became the last stopping place for one of the university's most restless personalities. Psychologist Edwin G. Boring, removed from the turmoil outside that had erupted with the assassinations of Martin Luther King, Jr., and Robert F. Kennedy in the spring, lay in bed, his once massive frame crumbling under its own weight as the fatal myeloma dissolved his bones. He was alternately lucid and delirious, as his friends and colleagues and former students came by to pay their last respects to one of the architects of American psychology and "perhaps the last great universalist of the profession."[5]

Boring's enormous energy was finally waning after a career that spanned six decades, including a quarter-century at the helm of the Harvard psychology laboratory. He was one of the last leaders of his generation, a generation that had come of age during the First World War and guided the professional fortunes of psychologists through the boom times of the 1920s, the lean years of the depression, and the era of lavish growth following World War II. There was now a veritable army of psychologists in America, nearly 30,000 strong, more than a hundred times as many as when Boring finished graduate school. Now, as psychology had reached an apogee in its remarkable trajectory in American culture, Boring was leaving the scene, his own life and work revealing in microcosm the forces and events that had shaped the field he loved.

Boring had begun his life in psychology as a researcher, and he never lost his devotion to the laboratory even after he abandoned its apparatus in favor of the writing table and the typewriter. In the 1920s he sought to defend the experimental tradition in psychology and its quest for a nomothetic science of universal laws by taking up the unlikely weapon of history. His particularistic, idiographic approach to the "great men" of psychology and their work was mobilized in support of an account of cumulative scientific advance. Thus his intensely personalistic view of history was held in dynamic tension with his faith in the inexorable progress of science.

Professional success did little to assuage Boring's chronic sense of personal failure, and in the 1930s he turned to psychoanalysis for relief, despite his scientific skepticism. World War II brought opportunities for practical service that he could not ignore, and he willingly lent his energies to the tasks of explaining and facilitating the usefulness of psychology in the war effort. Finally, after the war, Boring seemed to acquire the balance and maturity that he had been seeking for years, and he comfortably assumed the mantle of an elder statesman.

5 S. S. Stevens, "Edwin Garrigues Boring: 1886–1968," *American Journal of Psychology,* 1968, *81:*589–606, quote on p. 605.

During the last two decades of his life Boring's professional networks continued to proliferate as psychology grew with unprecedented vigor. As early as 1954
he predicted that, if the current rate of growth continued, the number of psychologists would equal the entire world population by the year 2100.[6] Although such
an outcome was clearly absurd, it did dramatize the enormous expansion of the
field. Perhaps a future full of psychologists – even if the majority were pursuing
applied psychology – was preferable to the alternative.

Edwin Boring passed away on 1 July 1968 without ever having set foot in Germany. Death finally separated him from the scientific fatherland that had nurtured
his early career and provided a backdrop to his continuing quest for personal and
professional maturity. In keeping with his last wishes, his body was cremated and
the ashes scattered in Mount Auburn cemetery in Cambridge.

A year after his death, at the 1969 annual meeting of the American Psychological Association, a session was held "In Memory of Edwin G. Boring." There
friends and colleagues remembered E.G.B. and reflected on his outsize personality and his complex legacy.[7] The 1969 meeting itself was replete with reminders
of the great man who had helped bring psychology to its present pass. Even the
president's address by George Miller, "Giving Psychology Away," could be read
as an epitome of Boring's lifelong struggle to help psychological science achieve
cognitive and cultural authority, presented by an erstwhile student and colleague.
The man whose life personified the rise of a technoscientific profession in America was gone. Now others would mark the place of "psychology's Boswell" in
history.[8]

6 Boring, quoted in Fillmore H. Sanford, "Across the Secretary's Desk," *American Psy-
  chologist,* 1954, *9:*125.
7 The session was chaired by Julian Jaynes, with remarks by Harry Helson, Henry Murray, Saul Rosenzweig, and Gardner Murphy. For published versions, see Harry Helson,
  "E.G.B.: Early Years and Change of Course," *American Psychologist,* 1970, *25:*625–629;
  Julian Jaynes, "Edwin Garrigues Boring: 1886–1968," *Journal of the History of the
  Behavioral Sciences,* 1969, *5:*99–112; Saul Rosenzweig, "E. G. Boring and the *Zeitgeist:
  Eruditione Gesta Beavit,*" *Journal of Psychology,* 1970, *75:*59–71.
8 Donald T. Campbell used the label in his 1977 William James Lectures at Harvard;
  EGB/4229.85.

Epilogue: Science in Search of Self

The postwar alliance for scientific professionalism began to peak around 1970 as the distinctive contours of the psychology community began to merge with the cultural landscape of America. A seemingly limitless market for psychological expertise enabled the profession to continue its remarkable expansion, with no end in sight. In 1970 membership in the American Psychological Association reached more than 30,000, an increase of nearly a hundred-fold over the span of a single professional lifetime.

By this time scientific interests and social concerns had coalesced around the study of personality and the concept of the self. Interest in personality theory was high among academic psychologists, reinforced by the "golden age" in the popularization of psychoanalytic ideas in the postwar period. Possessing vast interpretive flexibility, the self was constructed as an object of research and a locus of personal and cultural concern. After disappearing for decades from the psychological literature, theoretical and empirical studies of the self proliferated after the war. Between 1940 and 1970 some two thousand publications related to the concept of the self appeared.[1]

The cultural preoccupation with the self was already well established when writer Tom Wolfe christened the 1970s the "Me Decade." In a series of brilliant essays he captured the widespread and intense self-absorption of the American middle class that was associated with sexual liberation, recreational drug use, psychotherapeutic and religious cults, and various other forms of self-indulgence. Wolfe likened this new preoccupation with the self and individual experience to the "great awakenings"

---

1 For a useful overview, see Kenneth J. Gergen, *The Concept of the Self* (New York: Holt, Rinehart and Winston, 1971), which tallied more than 2,000 published studies in self-psychology since 1940. See also John C. Burnham, "Historical Background for the Study of Personality," in E. F. Borgatta and W. W. Lambert, eds., *Handbook of Personality Theory and Research* (Chicago: Rand McNally, 1968), pp. 3–81.

of religious feeling that had occurred in the eighteenth and nineteenth centuries. This "Third Great Awakening" was a product of the mid-twentieth century economic prosperity and global political dominance enjoyed by the United States, which provided the means for its citizens to pursue personal enlightenment on a mass scale. Wolfe used the metaphor of alchemy to describe the heart of the Me Decade: "The old alchemical dream was changing base metals into gold. The new alchemical dream is: changing one's personality – remaking, remodeling, elevating, and polishing one's very self . . . and observing, studying, and doting on it. (Me!)"[2] This new alchemy of the self derived many of its formulas and working methods from the postwar expansion of psychology as a helping profession.

A less cynical view of the postmodern psyche was presented by psychiatrist Robert J. Lifton in his 1968 essay "Protean Man." The product of particular cultural forces and historical events acting on human nature, the essay described a psychological propensity toward a continual process of self-definition and redefinition. Influenced by David Riesman and Erik Erikson, Lifton sought to account for selfhood as a flexible and dynamic response to the environment. His studies of the survivors of Hiroshima had revealed to him the processes of personal self-renewal and re-creation as they took place under extreme circumstances.[3] In the modern world, change appeared to be the only constant of personality.

The metaphorical language of alchemy and the mythological reference to Proteus both invoked the transformative thrust of modern psychology. Qualified by its claims to scientific credibility, psychology had become a discipline devoted to helping people cope with their world. In an era that brought the burden of atomic warfare as well as unparalleled prosperity to Americans, it was perhaps not surprising that the management of human relations emerged as a central problem.

In postwar America, psychology prospered by becoming a protean profession, responsive to the needs of any and all. As the pursuit of cognitive authority became bound up with the exercise of expertise, it became a profession driven by the demands of the marketplace. Psychological knowledge became a cultural commodity that was easily manufactured and widely distributed by a self-sustaining community of technoscientific professionals.[4]

2 Tom Wolfe, "The Me Decade and the Third Great Awakening," in *Mauve Gloves and Madmen, Clutter and Vine* (New York: Bantam, 1977; orig. pub. 1976), pp. 111–147, quote on p. 126. See also Christopher Lasch, *The Culture of Narcissism: Life in an Age of Diminishing Expectations* (New York: Norton, 1978).

3 Robert Jay Lifton, "Protean Man," in *History and Human Survival* (New York: Random House, 1970; orig. pub. 1968), pp. 311–331. See also Lifton, *The Protean Self: Human Resilience in an Age of Fragmentation* (New York: Basic Books, 1993).

4 Ernest Gellner's observations, paraphrased by Nathan Hale, on the social mission of psychoanalysis as "the cure of souls under the aegis of science" also apply to academic psychology. Nathan G. Hale, Jr., *The Rise and Crisis of Psychoanalysis in the United States: Freud and the Americans, 1917–1985* (New York/Oxford: Oxford University

More than a century ago, in 1891, Herbert Nichols was awarded a Ph.D. in psychology at Clark University, becoming the first in a long line of students who earned their degrees there under G. Stanley Hall. The fledgling psychologist landed a position as an instructor at Harvard, where he assisted Hugo Münsterberg in the newly expanded psychology laboratory. Confident about the future of psychology at Harvard and elsewhere, Nichols boldly predicted that "the twentieth century will be to mental science what the sixteenth century was to physical science, and the central field of its development is likely to be America."[5] A few years earlier his other senior colleague, William James, had expressed a similar faith in experimental research but preferred to "psychologize" about the mind on the basis of personal experience. His insights informed his classic *Principles of Psychology* (1890), which included a chapter on "The Consciousness of Self." Rejecting the notion of a transcendental self, James discussed the empirical self, or what each person "is tempted to call by the name of *me*."[6] A fundamental and enduring interest in self-discovery animated psychology for more than a century and provided the basis for the rise of a major technoscientific profession. In the closing years of the twentieth century, American psychology remained headed in the direction of self-discovery, even if there was no ultimate destination in mind.

Press, 1995), p. 380. See also Ernest Gellner, *The Psychoanalytic Movement: The Cunning of Unreason* (Evanston, Ill.: Northwestern University Press, 1996; orig. pub. 1985).

5 Herbert Nichols, "The Psychological Laboratory at Harvard," *McClure's*, 1893, *1:*399–409, quote on p. 409.
6 William James, *The Principles of Psychology*, vol. 1 (Boston: Henry Holt, 1890), p. 291.

Index

Abel, T. W., 61, 81
Achilles, P. S., 61
Adams, D., 129
Adams, E. S., 39
Adams, G., 34
Adams-Webber, J., 38
Adjutants General's Office, 39, 41, 42, 55, 99, 100, 101, 138, 139, 172
Advisory Board on Clinical Psychology, 139, 140, 172
Albee, G. W., 246
Allport, Gordon W., 61, 66, 67, 68, 82, 115, 116, 117, 118, 119, 120, 126, 155, 163, 189, 191, 212, 213, 214, 221
Almond, G., 124
Alpert, R., 250
American Association for the Advancement of Science, 18, 46, 61, 67, 126, 244
American Association for Applied Psychology, 31, 36, 40, 41, 44, 45, 46, 55, 58–70, 71, 73, 85, 89, 93, 128, 130, 131, 132, 133, 134, 135, 136, 139
American Chemical Society, 62
American Council of Learned Societies, 117
American Council on Education, 117
American Men of Science, 18, 21, 30, 77
American Philosophical Society, 48
American Psychological Association, 1, 10, 13, 16, 18, 21, 22, 29, 30, 31, 36, 37, 40, 41, 44, 45, 46, 52, 55, 57, 58–70, 71, 73, 85, 89, 116, 119, 126, 127, 130, 131, 132, 133, 136, 137, 139, 141, 161, 162, 163, 164, 165, 173, 174, 177, 179, 198, 204, 211, 212, 219, 229, 231, 234, 239, 243, 244, 245, 246, 247, 251, 254, 255, 256, 257, 259, 262, 263
American Psychologist, 87, 88

American Teachers Association, 60, 61, 66
Ames, L. B., 165
Anastasi, A., 164
Anderson, E. E., 70
Anderson, F. J., 123
Anderson, J. E., 27, 61, 67, 68, 70, 136
Anderson, W., 120
Andrews, T. G., 84
Angell, J. B., 19
Angell, J. R., 17, 18, 19, 39, 43, 44, 45, 117
Ansbacher, H., 49
Appelbaum, S. A., 230
Applied Mathematics Panel, 53, 144, 170
Applied Psychology Panel, 51–54, 144, 146, 161
Archibald, K., 180
Arensberg, C., 124
Armstrong, C. P., 73, 77, 78, 83
Army General Classification Test, 99, 101, 102, 103, 104, 107
Army Specialized Training Program, 103
Arnold, H. H., 169
Arnold, W. J., 224
Ash, M. G., 7, 26
Association for Women Psychologists, 165
Atomic Energy Commission, 176
Auden, W. H., 124
Aviation Cadet Qualifying Examination, 108
Aviation Psychology Program, 107, 108, 109, 110

Baba Ram Dass, 250
Back, K. W., 226, 250
Baker, R. R., 173
Bannister, D., 230
Barclay, L., 74
Barnard, C. I., 188